Fritz Dross, Birgit Nemec, Igor Kąkolewski (eds.)
Sexuality, Family Planning, and Reproduction

Historical Gender Studies | 11

Editorial

Research in historical gender studies began as an endeavour to acknowledge women's contributions to history, to make their voices heard. Today, scholars of historical gender studies work on various topics, such as the history of gender relations, the history of feminism, of discourses of masculinity, as well as the history of the everyday lives, the persecution, the disfranchisement, and the struggles of LGBTQI* people. The book series **Historical Gender Studies** offers a forum for innovative research on the many facets of this field to create a collaborative project for scholars of history, gender studies, cultural studies, and others.

Fritz Dross, born in 1965, works as an assistant professor at the institute for the history of medicine and medical ethics at Friedrich-Alexander Universität Erlangen-Nürnberg in Erlangen. The historian did his doctorate on the history of urban hospitals around 1800 in Düsseldorf. In his research, he focuses on the history of hospitals, health care provision and poor relief in the early modern period as well as the history of gynecology in the 20th century.
Birgit Nemec is a professor for the history of medicine and the biomedical sciences at University of Vienna, Austria. She is interested in the roles of patients and activists in the negotiation of knowledge and practices in the new history of reproduction. Birgit Nemec is member of the Young Academy of Science.
Igor Kąkolewski, born in 1963, is a historian and director of the Center for Historical Research Berlin of the Polish Academy of Sciences. His research focuses on the early modern history of the Central-Eastern Europe and the cultural history of mental disorders of the ruling elites in Europe in the 16th and 17th centuries.

Fritz Dross, Birgit Nemec, Igor Kąkolewski (eds.)

Sexuality, Family Planning, and Reproduction

Historical Dimensions in Central and Eastern Europe from 1600 until Today

[transcript]

Bibliographic information published by the Deutsche Nationalbibliothek
The Deutsche Nationalbibliothek lists this publication in the Deutsche Nationalbibliografie; detailed bibliographic data are available in the Internet at https://dnb.dnb.de

This work is licensed under the Creative Commons License BY-NC-ND 4.0. For the full license terms, please visit the URL https://creativecommons.org/licenses/by-nc-nd/4.0/deed.de.
Creative Commons license terms for re-use do not apply to any content (such as graphs, figures, photos, excerpts, etc.) not original to the Open Access publication and further permission may be required from the rights holder. The obligation to research and clear permission lies solely with the party re-using the material.

2025 © Fritz Dross, Birgit Nemec, Igor Kąkolewski (eds.)

transcript Verlag | Hermannstraße 26 | D-33602 Bielefeld | live@transcript-verlag.de

Cover design: Kordula Röckenhaus
Cover illustration: Historische Bildpostkarten – Universität Osnabrück – Sammlung Jutta Assel, www.bildpostkarten.uos.de
Printing: Elanders Waiblingen GmbH, Waiblingen
https://doi.org/10.14361/9783839470831
Print-ISBN: 978-3-8376-7083-7 | PDF-ISBN: 978-3-8394-7083-1
ISSN of series: 2627-1907 | eISSN of series: 2703-0512

Printed on permanent acid-free text paper.

Contents

Introduction 7

The Architecture of Sexuality
The Customs of the Polish Nobility and its Influence on Architecture During the Early Modern Period
Aleksandra Jakóbczyk-Gola 13

City Midwives in Thorn and Danzig
Hebammeneide and *Hebammenordnungen* in 17th- and 18th-century Polish Prussia
Katarzyna Pękacka-Falkowska 29

From Sex-Driven Maids to Population Regulation to the Creation of the Housewife
Reproductive Struggles in Saxony and the Habsburg Empire in the 18th Century
Tim Rütten 45

Is marriage so Sacred?
Extramarital Births in West Prussia circa 1900
Hadrian Ciechanowski 57

The *Jewish Women's League* of Breslau
Its Efforts to Protect Reproductive Health and the Health of Women and their Children in the Early 20th century
Izabela Spielvogel 77

Debating Birth Control in Interwar Polish-Jewish Contexts
Ewa's Commitment to the Shaping of a Modern Jewish Polish Family Image
Heidi Hein-Kircher 95

"From Girls into Women, from Boys into Men"
An Expert's Discourse and the Press in a Medium-Sized City in Interwar Poland.
The example of Tarnów
Marcin Wilk 107

Divergent Narratives on Family Planning in Interwar Poland
Between "secret marriage tricks" and "the obligation of maternity"
Elisa-Maria Hiemer 119

Single Mothers and the Issue of Motherhood in Essays and Popular Cinema in Poland in the 1930s
Małgorzata Radkiewicz 133

"War of births"
Midwifery under German Occupation in the Wartheland, 1939–1945
Wiebke Lisner 149

Continuity of "Race Hygiene"?
Discourses and Practices of Sterilization in the Soviet Occupation Zone and the Early GDR
Stefan Jehne 167

Physicians as the Main Actors in the Debate over Birth Control in Czechoslovakia, 1920s–1960s
Veronika Lacinová Najmanová 181

"Killing of Unborn Children" and "Pornography"
Discourses on Sexuality and Reproductive Rights in Post-war Poland
Michael Zok 197

Authors 211

Introduction

On Friday 30 January 2020, it was not only in Geneva that things got heated. Thomas Beddies of the *Institute for the History of Medicine and Ethics in Medicine* at the *Charité Berlin*, Igor Kąkolewski, Director of the *Centre for Historical Research Berlin* of the *Polish Academy of Sciences* (CBH PAN), and Fritz Dross, then Chairman of the *German-Polish Society for the History of Medicine*, met in Berlin to discuss the joint preparation and organisation of a conference on the history of reproduction. In the course of the meeting, they agreed to jointly prepare a conference entitled *Mother, Father, Child – The History of Reproduction* to be held in the premises of the CBH PAN in June 2021, deviating from the usual date of the autumn congress of the *German-Polish Society for the History of Medicine*.

During the Berlin meeting, the *WHO* in Geneva declared a public health emergency of international concern. On 22 January, Lothar Wieler, director of Berlin's *Robert Koch Institute*, publicly reassured the German news programme Tagesschau that "only a few people can be infected by others".[1] But the situation changed faster than any prediction – the rest of the pandemic story does not need to be told here. While the Call for Papers, finally published in July,[2] was still being finalised, the entire event had to be converted to an online format, for which there was little experience in the first half of 2020. At the same time, the circle of responsible organisers expanded to include Birgit Nemec, who took up a professorship at the Berlin Institute for the *History of Medicine and Ethics in Medicine* at the *Charité* in Berlin in April 2020.

In June 2021, a total of 27 papers by speakers from Poland, Germany, the Czech Republic, Austria, Denmark and the USA presented a wide range of historical approaches to the topic, structured into sections on 'Reproductive Be-

---

1   https://www.merkur.de/deutschland/corona-rki-robert-koch-institut-hopkins-zahlen-infektionen-statistik-kritik-wieler-deutschland-zr-13602916.html (05.06.2025).
2   https://www.hsozkult.de/event/id/event-92940, (05.06.2025).

haviour and the Private: Numbers and Meanings', 'Mother and Child', 'Public Health and Public Discourse in the Interwar Period', 'Experts and the Public in Discourses on Reproduction', 'Reproduction and the Material World: Architecture and Industrial Design', 'Midwives as Experts', 'Clerical, Political and Medical Advice', 'Abortion Cultures', and finally, 'Silent and Noisy Revolutions: Discourses of Reproduction in the Late 20th Century'. During the conference it became abundantly clear that this issue, which is still topical and will continue to be so in the future, requires more historical depth than ever before.

Reproduction is rarely out of the news. It is an issue in ongoing debates about assisted reproduction and old and new concepts of the family, or violence during childbirth, and has socio-cultural implications in terms of medical advances such as uterine transplants, not to mention the intense debate about abortion that has been going on for a century. To take just one example, the role of women's participation in the parliamentary elections in Poland in October 2023 can be highlighted, not least because of the previously intensified debate on abortion and reproductive rights. The debate has flared up again in several other national contexts, not only but also in Central European countries, bringing into the media and political arena an issue that is particularly important from a historical perspective because it raises fundamental questions about medical paternalism, autonomy, medical responsibility, the availability and safety of medical products and services, and the distribution of knowledge and ignorance.

Questions about the socio-political environment of reproduction and individual rights have been debated for decades. While public debate has intensified on the western side of the Iron Curtain since the formation of critical social groups in the 1970s, we still need more historical analysis of how the economic, social and cultural changes since the late 1980s have affected the eastern side. For the West Berlin collective *Brot und Rosen* in 1971, for example, industrial products such as the first 'pill', Anovlar, stood for the harmful collaboration of politics, doctors, industry and the church in establishing authority over the female body. Criticism and alternative, personal perceptions and interpretations of industrial hormones were vociferously expressed in various spaces and formats. From a historiographical perspective, these voices contribute to an ambivalent picture of reproduction after the 'sexual revolution', shaped by conflicting evaluations and viewpoints, a topic to which recent research has turned. But how did this work in the socialist countries of Central Europe?

The aim of the conference and of this volume is thus to explore the historical dimensions of these problems in a broad field where human biology, reproductive medicine, family policy and state social programmes intersect with fundamental conceptions of desired or feared social developments projected onto religious and cultural ideals. Taking the changing political, social, cultural and scientific relations between Germans and Poles and the corresponding interconnections in Central Europe as an example, the historical understanding of the role of medicine in conceptions of family and gender, as well as the role of relevant socio-cultural institutions and medical development professionals, will be examined. The history of reproduction opens the door to fundamental questions of historical anthropology.

The history of reproduction is a timely topic and an emerging field of research at the intersection of history, medical history, anthropology, STS, media and gender studies, and many other disciplines. Interdisciplinary collaborations, such as the *Strategic Research Initiative Reproduction* at the *University of Cambridge*, have shown that, unlike the academic history of pregnancy and childbirth, questions now encompass multiple spheres of life, from contraception to cloning and populations, and focus on their ongoing renegotiation. In light of the debates outlined above, it can be argued that reproduction is currently emerging as a rapidly growing interdisciplinary field of research, raising complex questions that are being answered in a variety of settings, including museum collections, testimony seminars and oral history labs. This is because, as Lauren Kassell, Nick Hopwood, and Rebecca Flemming have argued in the seminal reader *Reproduction – from Antiquity to the Present* (2018), reproduction is a subject that "has such as wide scope, from the most intimate experiences to planetary policy, and because it raises such large and difficult questions", with rapid technological innovations: IVF, PGT, NIPT, genome editing – because "innovation fuels controversies over science and technology, economics and politics, ethics and religion while children keep on being born".

And yet, in the face of these old, renewed and new interests of individuals, politicians, scientists and scholars in the history of reproduction, in the attempt to understand historical and current practices, sometimes with the aim of searching for new directions, a volume focusing on a Central and Eastern European perspective was still a desideratum. This is why we are incredibly excited about this volume and the wonderful group of scholars it brings together for the first time.

The production of this volume has also been repeatedly delayed by the consequences of the pandemic and epidemiological restrictions on cross-border scientific work. We have finally decided to present the 13 essays in chronological order. Aleksandra Jakóbczyk-Gola begins with an article on the early modern period, dealing with the question of the concrete place of sexuality, especially in the architectural debate (The Architecture of Sexuality). Katarzyna Pękacka-Falkowska (City midwives in Toruń and Gdańsk) deals with the regulation of midwifery in urban contexts in the 17th and 18th centuries, which has so far been neglected in both Polish and German historiography. Another article on the early modern period was written by Tim Rütten and analyses the idealised gender role of housemaids on the way to the creation of the bourgeois housewife (From Sex-Driven Maids to Population Regulation to the Creation of the Housewife). After a major chronological leap to the turn of the century around 1900, Hadrian Ciechanowski examines the population statistics behind the debates on illegitimate births in Prussia (Is marriage so sacred?). Izabela Spielvogel then looks at the activities of a religiously based Jewish women's association in the field of pregnancy and child health (The Jewish Women's League of Breslau). Heidi Hein-Kircher analyses the role of the abortion debate in the Jewish weekly *Ewa* in the interwar period (Debating Birth Control in Interwar Polish-Jewish Contexts), while Marcin Wilk also analyses the role of the press in the establishment of a medical expert position in a medium-sized Polish town (From Girls to Women, from Boys to Men – The Example of Tarnów), Elisa-Maria Hiemer analyses the different and contradictory narratives on family planning in interwar Poland (Divergent Narratives on Family Planning in Interwar Poland), and Małgorzata Radkiewicz examines the representation of single mothers and their role as mothers in Polish films of the 1930s (Single mothers and the issue of motherhood in essays and popular cinema in Poland in the 1930s).

Wiebke Lisner looks at midwives under German occupation in 'Wartheland' (Midwifery under German occupation in Wartheland, 1939–1945), focusing in particular on the racist discourse about 'German' and 'Polish' midwives in a 'war of births'. Stefan Jehne's study focuses on the period after the Second World War and traces the continuity of racial hygiene in the Soviet occupation zone and the early GDR using the example of sterilisation practices (Continuity of 'racial hygiene'?). In an article covering a longer period Veronika Lacinová Najmanová examines the expert role of female, but especially male, doctors in the changing debate on contraceptive practices from early Czechoslovakia after the First World War to the end of the formation phase of socialist Czechoslovakia (Physicians as the Main Actors in the Debate over Birth Control in Czechoslovakia, 1920s-1960s). Finally, Michael Zok analyses the Polish postwar debate, particularly on abortion, from the immediate post-war period in the Socialist People's Republic and its crises to the Third Republic in the 1990s (Discourses on Sexuality and Reproductive Rights in Post-war Poland).

It is a great relief to see this volume in print after many years of preparation! Many colleagues have been involved and we owe them a debt of gratitude. First of all, of course, the authors, including those who submitted manuscripts but who, for understandable reasons, were unable to complete the process with us, and especially those who were not deterred by the difficulties of the production process and whose important research is now available in this volume. Finally, the willingness of the *transcript* publishing house, namely Mirjam Galley and Annika Linnemann, to include this volume in the *Historical Gender Studies* series and to provide us with professional support throughout the long road to the book was central and indispensable. We would like to thank the *Archiv Historische Bildpostkarten* at the *University of Osnabrück* for providing the cover illustration. Luisa Bott prepared the manuscript for typesetting.

Small teams at the *Institute for the History of Medicine and Medical Ethics* at the *FAU Erlangen-Nuremberg*, the *Institute for the History of Medicine and Ethics in Medicine* at the *Charité Universitätsmedizin Berlin*, and the *Centre for Historical Research Berlin* of the *Polish Academy of Sciences* (CBH PAN) were involved in the work on this volume and the conference that preceded it. In Erlangen, Meredith Alongi proofread and corrected all the manuscripts and provided essential editorial advice. We would also like to thank Maximilian Welsch, János Hübschmann (in Erlangen) and Philipp Gutsche (in Berlin) for their help in locating bibliographical references. Renate Rittner handled all financial matters with her usual stoic calm and unflappable good humour.

The book was supported by the *Arts and Humanities Research Council* and the *German Research Foundation* through a UK-German Funding Initiative in the *Humanities Grant* (NE 2419/1-1) and by the *German Federal Ministry of Education and Research* and the *German Aerospace Centre* (01UL1907X).

The preparation of the grant application to the *German-Polish Science Foundation*, the organisation of the conference and the work on the volume were largely carried out by the staff of the *Centre for Historical Research Berlin* of the *Polish Academy of Sciences*: Dr Maciej Gugała, Dr Milena Woźniak-Koch and Anna Kuc, to whom the editors would like to express their gratitude for their work and support.

Finally, we are very grateful to the *German-Polish Science Foundation*, which, not for the first time, has financially supported a conference and the subsequent publication project of the *German-Polish Society for the History of Medicine*, thus making it possible in the first place.

The Architecture of Sexuality
The Customs of the Polish Nobility and its Influence on Architecture During the Early Modern Period

Aleksandra Jakóbczyk-Gola[1]

Abstract *This article is dedicated to the question of sexuality and family relations among nobility in Early Modern Poland and their expression in architecture. The author analyses old architectural treatises to find some information about representing the role of men and women, the interpretation of love and customs connected with sexuality in Polish manor houses. The family model in Early Modern Poland differed significantly from the traditional one.*

Sexuality in the old Polish culture, especially of the nobility is still an interesting topic, not only for research. Family patterns have changed, but they are constantly inspiring, because it is an issue close to everyone. It is easy to identify with members of old families, one can understand their needs, emotions, ambitions. A closer look at past families also allows to better define your own position and better understand the relationships between family members in today's world. Importantly – the family, and its most intimate spheres, such as sex and reproduction, were and still are connected to the home, the place where they are present, which creates a safe space for them. It is therefore worth taking a look at how these two concepts – architecture and sexuality – were related to each other in early modern Poland.

Polish architectural writing of this period is quite extensive. Interestingly, these are often practical guides, which were intended to help landowners not so much in erecting a house themselves, but in seeing to its construction, in deciding where to build it, in planning its rooms. Many times these treatises have

---

1   University of Warsaw / Polish History Museum.

already been described in terms of their architectural achievements, new ideas and construction techniques.[2] What is missing, however, is a look at the house as a living space. Its layout is the realization of the relationships that prevailed in the family, the role that women and men played in the society in the old Polish culture. Already in this period, between the end of the 16th century and the 18th century, we can observe clear changes in architecture resulting from the transformation of these relationships.

Therefore, the question should be asked, to what extent was the architecture of the Polish manor a reflection of the stereotype of certain sexual behaviors? How were intimate spaces created, above all the common bedroom of the spouses, built far from the main representative rooms. Where were the children's and parents' rooms located, and in what relationships did they remain?

However, in order to properly interpret the meaning of the rooms, it is worth first looking at the concept of love in old Polish Commonwealth culture. Understand how spiritual intimacy was interpreted, on what principles marital relationships were built. The love in early modern times in Poland is an unusual mixture of casual enough customs, vitality and joy of the body, with imposed, resulting from Christian morality, strictures. Unique to Polish culture in the context of Europe at the time was the rather strong intimacy of spouses and the distinct social position of women, especially widows.

Concept of Love

Sensuality in the Polish Commonwealth is a concept that developed in a state of tension and internal conflict. Catholic teachings exhorted purity, abstinence, and the denial of earthly desires. This approach was not commended solely in Church teachings and sermons, but was also present in lay literature, whose intent to offer clear ethical stances. The culture of 'Sarmatism' was an opposing force, characterized by vitality and expressive displays of emotion and dynamism. Sarmatism viewed traits such as virility, sexual availability and hot temperament as highly positive.

The concept of love in the cultural world of early modern Poland was multifaceted. Many renaissance writers, including the famous Polish poet, '*poeta doctus*' Jan Kochanowski, understood it to mean a strong passion, a 'burning heart', which was distinct and removed from human reason and will, and which was in

---

2   Małkiewicz, Architektury, 1976, p. 13. Cf. Mieszkowski, Traktatach, 1970.

itself ground for various choices and attitudes. The Italian ideal of love, which was introduced through poetry and references to the rhetoric of Petrarch, was based on excessive displays of admiration for the other sex; this approach did not take hold in the culture of early modern Poland.

Emotional attachment was described in Old Polish in other terms. It was interpreted to be a feeling more akin to sentimentalism, 'love of the heart', which was tender and affectionate and stemmed from human closeness. It could only exist within the framework of marriage and it could be achieved solely through sacrament.

The word 'miłość', literally meaning 'love', was understood mostly in terms of a sensual experience, Veneris' play or erotic love. The Polish Dictionary of Bogumił Linde written in the early 19th century recommends describing other desires as forms of love, for example the love of money, to emphasize a desiring nature.[3] There was widespread agreement that love could take a negative, condemnable shape and could lead to the rejection of social norms.[4] According to Sebastian Klonowiec, a Polish Renaissance poet, love is connected to desire and shameless frenzy which changes into lust and can lead to madness.[5] Early modern Polish culture was filled with impulses and innuendos. Conversation and jokes at the time tended to be much less restrained than they would be in later centuries, and often involved sexual themes. The sharing of lewd pictures and linguistic jokes full of sexual symbolism were common amusements.

Customs and Sexuality

It is possible that the more upfront approach to sexuality was partly influenced by the dietary customs of the Polish nobility in the early modern period. Excessive consumption of animal protein, and the large number of calories consumed every day, likely led to a heightened libido.[6] Openness towards phys-

---

3   Kuchowicz, Miłość staropolska, 1982, p. 17.
4   Ibid. p. 19.
5   Ibid.
6   Żyromski, Nawyki żywieniowe, 2003, p. 102. While a peasant consumed only 3500 calories a day (including as much as 82% carbohydrates), a nobleman already consumed 5300 calories (78% carbohydrates), and a magnate as much as 6300 calories (only 70% of carbohydrates). The low carbohydrate content of the magnats' diet was due to the fact that they consumed a lot more protein through a meat-based diet, which provided calories, but also contributed to the popularity of many diseases, such as gout.

ical pleasures was also aided by alcohol which was consumed with little restraint. Drunkenness was a particularly masculine vice, and often used to embolden men in their contacts with women. Boredom was yet another reason for engaging in the pleasures of the flesh. The life of the land nobility followed a predictable routine; many journals and diaries from the time emphasize the monotony of life in the country and the limited, sometimes complete lack of entertainment.

It is also worth remembering that in the early modern period, people lived under constant threat of death, be it from natural causes, war, disease, or natural calamities. Sexuality provided a way of escaping the everyday fear connected with death. Because death seemed so inevitable, engaging in marriage was also often rushed. Erotic life in the early modern period was characterized by passion but also by harshness, brusqueness and sometimes even brutality. Eroticism had to be vivid in order to find place in a time when the threat of death was constant.

Another reason was the fear of loneliness – solitary life was not widely accepted. Marriages tended to last only 10–15 years, largely due to the high level of mortality during childbirth. After a spouse died, a new marriage was quickly arranged, sometimes within months, often based on personal choice, in contrast to first marriages which were usually arranged by the couple's parents.[7]

Progeniture – or, extending one's bloodline – was one of the most pressing reasons for wedding early. Wives were expected to be fertile. They were supposed to give birth to sons, who would then continue the works of their forebears. Daughters were not considered desirable because they joined other families and had to be provided with a dowry. Dowries were often a serious financial burden on the father or caretaker of the bride.

Marriage – Love and Sex

Marriage, therefore, offered a solution to numerous issues connected to sex and sexuality. It allowed the teachings and laws of religion to comingle with human nature itself and human sexual desire. Contemporary norms designated the home as the only place for passions of the flesh. Marriage unified the couple, through loyalty, friendship or deeper romantic feelings.[8] It also uni-

---

7 Lisak, Miłość staropolska, 2011, p. 66.
8 Kuchowicz, Miłość staropolska, 1982, p. 45.

fied the couple in a physical sense – Christian teachings saw marriage as the only acceptable way of extinguishing the fires of passion and did not consider sex within marriage as sin. However, even within the confines of marriage, the Church preached restraint, including abstinence on the many holy days of the Christian calendar.[9] Sermons often condemned excessive sexual activity. However, sex was also viewed as an important part of a functioning marriage. Sexual activity was the source of pleasure and a way of satisfying desires, and which later allowed for the realization of other goals and aspirations.

Sex was also considered a cure for many different kinds of illnesses, especially in virtue of the medical teachings of the time, which were based on the idea of the balance of the humors. Sexual intercourse was believed to help calm various phlegmatic or melancholic maladies, and to cure faintness.[10] Married life, or rather – sex – served only people of a certain temperament. People who were "bloodthirsty" or "moist", those in which these fluids, called, according to the interpretation of the writings of the ancient physician Galen – humors, prevailed. Likability was good primarily for sanguinarians, and was particularly harmful to melancholics.[11] Cold, dry and weak natures were not prepared for love and, it was advised, it was better for such to warm themselves by the stove than by the side of a wife.

Like all medication, it was supposed to be used responsibly and without excess, at the right time and in specific health conditions. Otherwise, intercourse could be harmful for the intestines or could lead to an imbalance of the humors. Jakub Kazimierz Haur, author of early modern encyclopaedias and economic guides for the nobility, advised that intercourse should be done in the morning on an empty stomach, or in the evening a few hours after lunch. Otherwise, too much activity could affect the stomach badly and interfere with digestion. Long rest after intercourse was also recommended, especially for women.[12] This concept of engaging in sexual activity for the sake of health and hygiene grew to be even more important in the Polish Commonwealth in the 18th century. It was believed that sex would lead to conception only if it was pleasurable and full of passion.[13]

---

9   Lisak, Miłość staropolska, 2011, p. 152–154.
10 Ibid., p. 163.
11 Kuchowicz, Obyczaje staropolski, 1975, p. 272.
12 Haur, Oekonomiey ziemianskiej, 1693, p. 200.
13 Lisak, Miłość staropolska, 2011, p. 163.

The Family Model in Early Modern Poland

The family model in early modern Poland differed significantly from the model propagated in the 19th century, which now is often referred to as 'traditional'. In the early modern period the husband was not necessarily the only source of income for the family, and the role of the wife was not limited to house chores. We have numerous examples of noblewomen who managed landed estates and had their own servants. This was not in contrast to the wishes of their husbands, but rather met with their full approval. Thanks to the energy and economic talents of their wives, noblemen could engage in what they viewed as more worthwhile duties: war and politics.[14]

Young noblewomen were married off in order for them to start a family, not in order to provide a livelihood. We have numerous surviving accounts of women who made their own decisions about who they would marry, rather than leaving the matter to their parents.[15] The betrothal period would typically be rather short, sometimes limited to just a few weeks if the couple was well matched in terms of social and financial status. Women were eager for the independence granted to them by marriage. The worry that a longer time might reveal faults in the lady's looks or character, also motivated short betrothal periods.[16] The betrothal period later began to expand, and in the 18th and 19th centuries it was not uncommon for it to last for years.

Widowed women had the unique position of being quite independent in early modern Poland. They were free to choose the next candidate for marriage and no one could force them into an arrangement they did not want.[17] Only in the 19th century did women lose the role of partners to their husbands and were relegated only to the bringing up of children. Eroticism became only an aspect of procreation.

Despite the fact that there was more balance between the sexes in early modern Poland, social expectations for men and women were very different. Noblemen were allowed to maintain sexual relations outside of marriage, as long as this did not harm family relations[18] (for example in terms of children's

---

14 Ibid. p. 131–132.
15 Ibid. p. 73.
16 Ibid. p. 60.
17 Ibid. p. 66.
18 Kuchowicz, Miłość staropolska, 1982, p. 447.

inheritance). Women on the other hand, were expected to be chaste and modest. Women therefore were supposed to behave in a reserved way and exhibit modesty and bashfulness. Mothers did not teach their daughters anything about sex and the 'duties of marriage' that they would come to face,[19] which meant that women would enter marriage ignorant of their own sexuality and the pleasures and dangers connected with it.

The Manor House as a Place for Love

The backdrop for the noble family in the Polish Commonwealth was the manor house. Its architecture served to express the cultural norms which outlined the roles of men and women and their relations between each other. The manor was usually a one-storey building with a pitched roof. Architectural theory shows that in the latter half of the 17th century there was a clear shift in the style of the architecture of the nobles' dwellings. An example of this shift can be seen in an anonymous work (sometimes attributed to Łukasz Opaliński, royal cavalry marshal) on the building of manors, palaces and castles according to the heavens and the Polish custom (*Krótka nauka budownicza* ...), which was published in 1659, and which is considered to be the earliest normative text on architecture written in Poland.[20] The treatise discusses the specifics of various rooms and the gradation of space within the house. The interior of the manor was divided into the official part, which was a large hall, and the representative dining room, which was the heart of every home in early modern Poland. To accommodate the dining room, Polish manors tended to be highest in the central section.

Rooms and Apartments

The intimate sphere of the house was composed of specialized rooms and apartments belonging to the Lord and Lady of the house. Their spatial location was supposed to help prevent non-marital sexual relations. The *Krótka nauka budownicza* as well as the appendix from the *Oekonomika*, an economic guide

---

19   Lisak, Miłość staropolska, 2011, p. 107.
20   N.N., Dworów, 1659.

written by Jakub Kazimierz Haur in 1679,[21] discuss separate apartments for the different sexes. These were composed of three or four interconnected rooms. The 'antechamber' was a semi-formal room for the reception of guests, came first. This room was also accessible by the servants. Next was the main room, where most social activities took place. Last was the bedroom (the 'retirata'), which functioned as a private room for resting during the day.

This spatial arrangement began to appear quite late in early modern Poland. In Europe it was popular since the Renaissance, but in the Polish-Lithuanian Commonwealth it became a feature of rich households around the mid-17th century.[22] This setup made clear that direction of movement within the house should begin in the ornamental rooms and more towards the more private sections of the house.

The day rooms of the Lady of the house were located near her husband's; the reason for this was, according to a number of treatises, to limit the danger of infidelity. A wife's infidelity was not followed by as dire consequences as in most of Europe, where, betrayed husbands had to duel with lovers or tried other methods of revenge – confiscation of property, public insult. On the other hand, men who had been cheated on by their wives (cuckold husbands) were subjected to widespread ridicule and tended to be the butt of many jokes and humorous poems.

The architecture and layout of noble dwellings made the infidelity of husbands easier. Late-baroque and rococo hunting palaces were usually constructed on a central plan which was referred to as 'molino da vento'. The origin of the word was connected to windmills because the layout resembled the four arms of a windmill. Such palaces were often the sites of romantic rendezvous, with each wing of the palace having a separate alcove. Hunting was considered a pastime appropriate only for men. The hunters, most of whom would have been married, were expected to keep each other's infidelities secret from wives and female companions, though the immodest behaviour of men during such trips was something of an open secret. Because of the cultural norms of this era, women did not treat their husband's romantic exploits seriously.

Another place commonly used for romantic escapades were public baths. We find discussion of the immoral behaviours in bathhouses already in medieval documents, which recommended the separation of the sexes in the baths. However, by the 18th century mixed bathhouses were common and were

---

21   Haur, Generalna Oekonomika, 1679.
22   Miłobędzki (ed.), Dworów, 1957, p. 63–64.

not considered anything out of the ordinary. Their architecture catered to the need for privacy; bathhouses in Warsaw, for example, boasted numerous separate rooms which provided convenient, and intimate, spaces for romantic adventures.[23]

Together or Separate?

Fig. 1: Architectural design of two manors, Piotr Świtkowski, Budowanie Wieyskie: Dziedzicom Dobr Y Possessorom toż wszystkim iakążkolwiek zwierzchność po wsiach i miasteczkach maiącym Do Uwagi Y Praktyki Podane. Z Figurami, publischer: Michał Gröll, Lwów, Warszawa 1782.

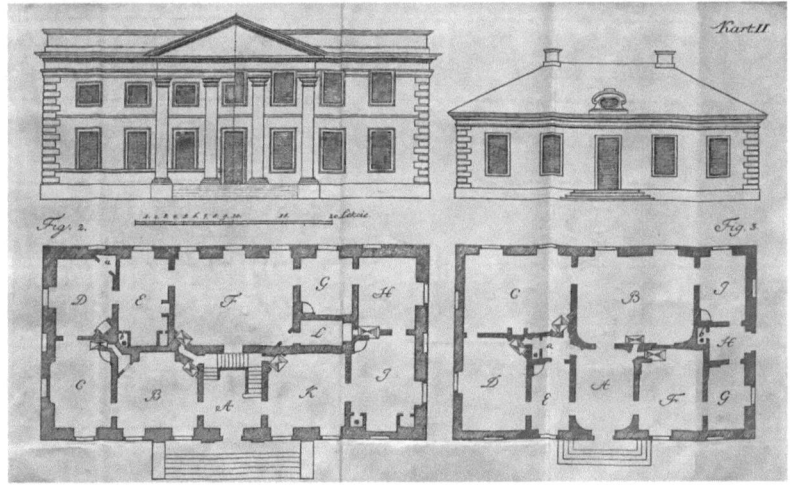

Let us return to the issue of the noble manor house – despite the relatively relaxed attitude towards husbands' infidelities, a married couple's love-life was considered very important. In early modern Poland it was usual for the husband and wife to share a bedroom, something very unusual in most of Western Europe.[24] The married couple's bedroom would be decorated either using romantic symbolism, or with depiction of naked couples in erotic scenes. Mytho-

---

23 Lisak, Miłość staropolska, 2011, p. 220.
24 Świtała-Cheda, Pomieszczeń, 2013, p. 160. Cf. Handley, Sleep, 2016.

logical figures such as Venus and Cupid were also common themes seen in decoration. The moral shift following the counter-Reformation forced many a landowner to remove such decorations from their bedroom.[25] Separate bedrooms for the spouses were considered to signal problems within the marriage.

The already-mentioned architectural treatise *Krótka nauka budownicza...* was intended, as the preface announces, to contain illustrations that would help in understanding all the instructions presented there. Unfortunately, these have not survived or were never actually drawn. In contrast, another early modern treatise features plans not only of manors, but also of farm buildings. Piotr Świtkowski's treatise published in 1782 is titled *Budowanie Wieyskie...* (Rural Building) and is intended for heirs and landowners.[26] Plate II shows the facades of two manors – a wealthier one on the left and a poorer one on the right (Fig. 1). On pages 105 to 107 the author of the treatise gives a detailed explanation of the letters that are on the plan. Let's start with the wealthier house. It represents the classic three-tract layout. The main axis is the vestibule (letter A) and the ballroom and dining room (letter F). The left side was designated for the rooms mainly of the lady of the house, but not only. There you can see the common bedroom for the couple – master and lady of the house (letter C) and the wife's daily room with a dressing room and small pharmacy (letter D). From it one could go to the children's room (letter E). On the other side of the axis are the rooms intended for the master of the house. There he has his additional bedroom (letter I), his study with a library (letter H), which, as Świtkowski writes, had to be away from the hustle and bustle, and his living room (letter G). The room marked with the letter K was also the couple's shared bedroom, but a winter one. It was undoubtedly better heated, as evidenced by the stove drawn on the plan.

The layout of the rooms in the poorer manor, depicted by Świtkowski more to the right of the illustration, is different. And in this case, the three-tract layout has been preserved, although not entirely symmetrical. The central part consists of a vestibule (letter A) and the so-called hall – a room for holding ceremonies and a parade dining room (letter B). Of course, it is much smaller than the one in the richer manor. To the right of these are the private rooms of the owners of the manor. They are connected by a common bedroom (letter G). From it you can go to one side to the lady of the house's room (letter F), and

---

25   Lisak, Miłość staropolska, 2011, p. 161.
26   Świtkowski, Budowanie Wieyskie, 1782.

to the other, through the dressing room (letter H), to the master of the house's study (letter I). The children live on the other side of the house (letter D). Their room is adjacent to the big room shared by all, a kind of living room and playroom (letter C), and on the other side to the dressing room and medicine cabinet – small pharmacy (letter E).

As can be seen from both of these plans, the architecture of the manor reinforced the previously described model of family relationships and dependencies in early modern culture in Poland. Also the model related to sexuality and physical contact. The couple used to spend the night together, but were separated by their daily duties and activities. If the house was more opulent, this separation was even more apparent – they both stayed in completely different parts of the house and their daily lives were concentrated there. Women were associated with children, music, play, a dressing room was necessary for them, as well as a medicine cabinet, home pharmacy – a room for storing and making various types of medicines. They dried herbs and cooked infusions there. The men's world, on the other hand, during the day was associated with intellectual work in the library, writing correspondence, managing property, compiling inventory books and records.

The devaluation of the significance of marriage began to take place in the 18th century. The husband and wife became almost strangers, and marriages became focused on material and social gain.[27] This was made evident in architecture as the bedrooms of the husband and wife were kept separate in this period, usually far away from each other. This allowed the husband and wife to pursue their own independent lives, including separate love lives. Intimacy and sexual relations were not considered an important aspect of marriage during the Enlightenment. Until the 18th century good relations between the spouses were manifested through harmonious and enduring sexual relations. A lack of physical intimacy was considered the end of the marriage. On the other hand, in the Enlightenment period husbands would often find influential lovers for their wives, hoping to use these relations to advance their own political careers. Also, noblewomen treated romance as an opportunity to exert influence or create lines of dependence, to pursue their own interests. Finally – it was simply fashionable. A refined lady of the late 18th century could not fail to have a lover.[28] In extreme cases the spouses would live in separate houses.[29]

27　Ibid. p. 128.
28　Kuchowicz, Obyczaje staropolskie, 1975, p. 300.
29　Lisak, Miłość staropolska, 2011, p. 128–129.

In some cases, especially in the houses of the royalty and the upper aristocracy, the living apartments included additional rooms fashionable at the time.[30] These were usually located in the corner sections of the buildings. This meant that the plan of the manor house changed. The chambers located out of reach of the guests of the house were all under women's control. Some of these chambers were highly elegant and stylish. The sitting rooms and children's rooms were associated only with women, as in the early modern period the upbringing of children was considered exclusively the domain of women. Similarly, music was considered a feminine activity. Rooms which were intended to be the location of family concerts were small and tended to be decorated in musical motifs. In early modern Poland, women were the ones who sang or played instruments and would meet in music rooms to play music together, while the husbands, brothers, and the rest of the family and guests could come to listen. Men themselves rarely played musical instruments.[31] A musical education was highly prized for young women, and a desired trait for wives. Young noblewomen therefore were often taught to enjoy music, sharpen their hearing, and play an instrument.

The architectural plan of the manor also allowed sole entry to the 'fraucymer' (from the German *Frauenzimmer*) from the suites belonging to the Lady of the house.[32] The fraucymer was where the court women of the Lady's house gathered. The women's court was divided hierarchically into the upper and lower 'fraucymer'. The upper fraucymer was the place for the guests of honour and the closest female companions of the Lady of the house, who often came from important noble families. It was also the place of education. The courts of rich women were often quasi-academies for girls and young women from the sphere of influence of the noblewoman. In the lower fraucymer were the ladies who served the upper-class ladies and the Lady of the house.[33]

Another very fashionable set of rooms, this time associated only with men, were cabinets. These were intended as places where a man could pursue his passions: study, reading, letter writing, and collecting of art, weapons or natural artefacts. These were the first museums in early modern Poland.

However, adding too many such fashionable rooms was criticized by the authors of architectural treatises. It was believed that more than three or four

---

30   Miłobędzki, Architektura, 1980, p. 72.
31   Bogucka, Białogłowa, 1998, p. 189.
32   Krótka nauka budownicza, 1659 p. 11.
33   More on the fraucymer in: Targosz, Sawantki, 1997, p. 47.

such rooms were superfluous.[34] The proliferation of such specialized rooms stemmed from the desire of the richer nobility to surround themselves in splendour, not from the need for comfort and practicality which were the cornerstones of architectural theory of this period.

The intimate life of people of early modern Poland was shaped by biology and constrained by the social and cultural norms of this time. Marriage was considered a social duty amongst the nobility and was often connected with material concerns. Architectural theory from this period aimed to express the social norms connected to marriage and procreation in the shape and style of the Polish manor house. The treatises outlined how to construct a home which was both comfortable, and in accordance with contemporary social norms. The treatises were written, and read, by men. It was men who were tasked with the creation of the proper topography for sexual and erotic encounters within the family. Men's role in the shaping, and maintaining, such norms was therefore much more dominant. Architecture became a tool for maintaining the permanence of traditions and customs of early modern Poland. The greatest expression of these traditions was the commonality of the married couple's bedchamber. As was pointed out by the mayor of Kazimierz, near Cracow, in his treatise *Stadło małżeńskie* written in 1561, such a union is "joint body, joint will, joint agreement, joint consent, joint concern and sadness, joint happiness, joint and equal freedom, common loss, common gain, common riches and poverty, equal dignity, and a shared bed for intercourse and rest, a commonality of all issues, works and dangers".[35]

References

Krotka Navka Bvdownicza Dworow, Pałacow, Zamkow. podług Nieba y zwyczaiu Polskiego, Kraków, u wdowy y Dziedzicow Andrzeia Piotrkowczyka, 1659.

BOGUCKA, Maria, Białogłowa w dawnej Polsce. kobieta w społeczeństwie polskim XVI-XVIII wieku na tle porównawczym, Warszawa, Trio, 1998.

HANDLEY, Sasha, Sleep in Early Modern England, Yale University Press, 2016.

---

34   Krótka nauka budownicza, 1659 p. 10.
35   Mrowiński, Stadło małżeńskie, 1561, Celichowski (ed.), 1890, p. 7.

HAUR, Jakub Kazimierz, Skład Abo Skarbiec Znakomitych Sekretow Oekonomiey Ziemianskiey. Na Polach [...] w [...] Krolestwie Polskim [...] Znaleziony, A za osobliwym Staraniem [...] Wykopany A oraz [...] Przykładami [...] Vbogacony, Wydany y do Druku Podany, Kraków, w Drukarni Mikołaia Alexandra Schedla [...], 1693.

HAUR, Jakub Kazimierz, Ziemianska Generalna Oekonomika. Obszernieyszym od przeszłey edycyey Stylem Svpplementowana I we wszytkich Punktach znacznie poprawiona [...] Punktami Partykularnemi, Interrogatoriami Gospodarskiemi, Praktyką Miesięczną, Modelluszami albo Tabułami y Frakcyami Arythmetycznemi, Obiasniona [...], Cracoviae, Typis Vniversitatis, 1679.

KUCHOWICZ, Zbigniew, Miłość staropolska. wzory – uczuciowość – obyczaje erotyczne XVI-XVIII wieku, Łódź, Łódzkie, 1982.

KUCHOWICZ, Zbigniew, Obyczaje staropolskie XVII-XVIII wieku, Łódź, Łódzkie, 1975.

LISAK, Agnieszka, Miłość staropolska. obyczaje, intrygi, skandale, Warszawa, Bellona, 2011.

MAŁKIEWICZ, Adam, Teoria architektury w nowożytnym piśmiennictwie polskim, Kraków, Warszawa, nakł. Uniwersytetu Jagiellońskiego, Państwowe Wydaw. Naukowe, (Zeszyty Naukowe Uniwersytetu Jagiellońskiego, Bd. 423), 1976.

MIESZKOWSKI, Zygmunt, Podstawowe problemy architektury w polskich traktatach. od połowy XVI do początku XIX w, Warszawa, Państwowe Wydaw. Naukowe, (Studia i Materiały do Teorii i Historii Architektury i Urbanistyki, Bd. 7), 1970.

MIŁOBĘDZKI, Adam (ed.), Krótka nauka budownicza dworów, pałaców, zamków podług nieba i zwyczaju polskiego, Wrocław, Zakład Narodowy im. Ossolińskich, (Teksty Źródłowe do Dziejów Teorii Sztuki, Bd. 7), 1957.

MIŁOBĘDZKI, Adam, GIEYSZTOR-MIŁOBĘDZKA, Elżbieta, MAŁKIEWICZ, Adam, JAWORSKA, Władysława Jadwiga, MOSSAKOWSKI, Stanisław, PIETRUSIŃSKI, Jerzy, RYSZKIEWICZ, Andrzej, STARZYŃSKI, Juliusz & TOMKIEWICZ, Władysław, Architektura polska XVII wieku. 1, Warszawa, Państwowe Naukowe, Sztuka polska XVII wieku, 1980.

MIŁOBĘDZKI, Adam, GIEYSZTOR-MIŁOBĘDZKA, Elżbieta, MAŁKIEWICZ, Adam, JAWORSKA, Władysława Jadwiga, MOSSAKOWSKI, Stanisław, PIETRUSIŃSKI, Jerzy, RYSZKIEWICZ, Andrzej, STARZYŃSKI, Juliusz & TOMKIEWICZ, Władysław, Architektura polska XVII wieku. 2. Album ilustracji, Warszawa, Państwowe Naukowe, Sztuka polska XVII wieku, 1980.

MROWIŃSKI, Jan, Jana Mrowińskiego Płoczywłosa Stadło małżeńskie, 1561, Kraków, nakł. Akademii Umiejętności, (Wydawnictwa Akademii Umiejętności w Krakowie. Bibljoteka Pisarzów Polskich, Bd. 12), 1890.

ŚWITAŁA-CHEDA, Mirosława, Nazwy pomieszczeń w szlacheckich dworach, in: Studia Łomżyńskie, (2013), p. 157–165.

ŚWITKOWSKI, Piotr, Budowanie Wieyskie. Dziedzicom Dobr Y Possessorom toż wszystkim iakążkolwiek zwierzchność po wsiach i miasteczkach maiącym Do Uwagi Y Praktyki Podane. Z Figurami, W Warszawie Y Lwowie, Nakładem i Drukiem Michała Grölla [...], 1782.

TARGOSZ, Karolina, Sawantki w Polsce XVII w. aspiracje intelektualne kobiet ze środowisk dworskich, Warszawa, Retro-Art, Rozprawy z Dziejów Nauki i Techniki, 1997.

ŻYROMSKI, Marek, Nawyki żywieniowe w dziejach rodziny polskiej, in: Roczniki Socjologii Rodziny, 15 (2003), p. 95–112.

City Midwives in Thorn and Danzig
Hebammeneide and Hebammenordnungen in 17th- and 18th-century Polish Prussia

Katarzyna Pękacka-Falkowska[1]

Abstract *This study explores the role of city midwives in early modern urban healthcare, examining their professionalization, regulation, and social status in Polish Prussia. It investigates municipal policies governing midwifery, training practices, and the evolving perception of childbirth assistance. By analyzing historical sources, the paper highlights the significance of midwives in shaping maternal care and public health in urban settings in early modern Thorn (Toruń) and Danzig (Gdańsk).*

Introduction

In the broader landscape of the history of reproduction, the exploration of midwifery in early modern Western Europe stands as a well-established research topic, offering rich insights into the practices and regulations surrounding childbirth. However, a notable gap persists in scholarly inquiry when it comes to the Polish-Lithuanian Commonwealth. While much attention has been directed towards the institutionalization of obstetrics and gynecology in the 19th and 20th centuries by Polish researchers,[2] the discourse surrounding midwifery in the region during the early modern period remains conspicuously unexplored. This gap not only challenges prevailing narratives in

---

1   Poznan University of Medical Sciences.
2   Cf. Stawiak-Ososińska, Położnicza, 2019. Stawiak-Ososińska, Kształcenia akuszerek, p. 157–171. Kurkowska, Akuszerka, p. 219–238. Mataniak, Akuszerki rządowe, 2017, p. 162–181. Kolka, Klinika Położniczo-Ginekologiczna, 2001. Kuźma-Markowska, Walka z "babkami", 2017, p. 189–215.

medical history but also perpetuates misconceptions regarding the regulatory frameworks for midwives in the early modern Poland-Lithuania.

The absence of comprehensive studies on midwifery in the Polish-Lithuanian Commonwealth has left a void in understanding of reproductive practices and healthcare dynamics in East-Central Europe. While some attention has been given to printed materials from the 16th and 17th centuries, offering glimpses into pregnancy, childbirth, and maternity care,[3] and to medical publishing of the late 18th century, when Polish medical nomenclature started to form,[4] the daily practices of midwives and the legal regulations governing their profession in back then Poland have remained largely unaddressed. Moreover, the neglect of early modern handwritten records by historians further exacerbates the perpetuation of stereotypes and myths regarding midwifery in these territories. One of them is the recurring opinion that in the Polish-Lithuanian Commonwealth there were no regulatory measures for midwives, which, in turn, were common in the West.[5]

This paper aims to address these research gaps by focusing on the role of city midwives (*Stadt-Hebammen*) in Thorn (present day Toruń) and Danzig (present day Gdańsk) during the late 17th and 18th centuries. By looking into archival records, it aims to shed light on the multifaceted responsibilities of sworn midwives (*vereidigte Hebammen*) and the evolving regulatory frameworks governing their practice. Through a comparative analysis of juridical documents from Thorn and Danzig, it seeks to reveal the nuanced evolutionary trajectory of midwifery in Polish Prussia, highlighting the progress towards formalization and professionalization witnessed in both cities. Furthermore, it aims to contextualize the gendered framework underpinning healthcare provision and (medicalized) policy of reproduction in German-speaking territories of back then Poland, wherein midwives, despite their essential roles

---

3   Węglorz, Poradniki medyczne, in: Justyniarska-Chojak, Konarska-Zimnicka (eds.), Per mulierem 2012, p. 412–432. Trzpiot, Stan in: ibid., p. 433–449. Justyniarska-Chojak, Troska, in: ibid., p. 407–420. Zaborowska, Pomoc przy porodach, in: Andrzej Karpiński (ed.), Wśród córek Eskulapa, 2009, p. 279–312.

4   Kocela, Sztuka babienia, 2020.

5   Cf. Wrześniewska, Bąk, Historia, 2012, p. 95. Matuszewska, Zarys historii, 2012, p. 49–50. Waszyński, Historia położnictwa, 2000, p. 56–58; Brzeziński (ed.), Historia medycyny, 1988, p. 463–464; Łapiński, Stan położnictwa, 1976, p. 1597–1599. Adamski, Rozwój położnictwa, in: Pielęgniarka i Położna 3/29 (1961), p. 14–15; and 4/29 (1961), p. 17–18.

in maternal and infant care, remained subordinate to male medical practitioners. By exploring the transition from character-based midwifery oaths to ordinances prioritizing medical skills and collaboration with city authorities, it also illuminates broader trends in medicalisation, moralization, and population policy.

The Organisation of Midwifery in the Polish-Lithuanian Commonwealth

In the Polish-Lithuanian Commonwealth, both unlicensed and licensed midwives offered their services to pregnant women. They also provided care and assistance during labor, delivery, and the postpartum period.[6] Nevertheless, only city-based sworn midwives formed the prominent female occupational group. Consequentially, they continuously straddled the line between the public and the private, being active in the large cities of former Poland. Shortly before the partitions of the late 18th century, the growing number of town councils in the Crown and the Grand Duchy of Lithuania began to draft specific regulations for licensing local midwifery. In the meantime, Polish Prussia, a very unique province of the Polish Crown, moved much more quickly than the other parts of the Commonwealth toward developing a regulated corps of midwives with specified professional duties and competencies.

In handwritten historical records from Polish Prussia, midwives working under oath (*vereidigte Hebammen*) started to appear at least from the second half of the 16th century. The first information about a licensed city midwife in Polish Prussia paid by the municipal authorities comes from the year 1581.[7] It was likely at that time that a city midwife office (*Stadt-Hebamme*) was established in Danzig, although as early as 1450 of one midwife working at a chamber for pregnant women in St. Elisabeth's hospital is mentioned in the sources.[8] In Thorn, in turn, sworn city midwives were contracted starting from

---

6   Kuklo, Badania nad historią kobiet, 2020, p. 26–27. https://doi.org/10.18778/0208-6050.107.02 (05.06.2025).
7   Pękacka-Falkowska, Arzneibuch, 2022, p. 187–189, 202–204.
8   Szarszewski, Proces medykalizacji, 2007, p. 193.

1601.[9] Nevertheless, as early as 1600, a plague midwife (*Pest-Hebamme*) was also employed.[10]

Over the next two centuries, city councils (*Stadträte*) of both cities put much effort into regulating and overseeing maternity care along with its female providers. Thus, the city councilors issued not only lists of duties for local midwives along with their oaths, but they also worked on comprehensive *Hebammenordnungen*. Such ordinances for city midwives along with their *Eiden* provide researchers with numerous pieces of information about how town councilors, municipal physicians, surgeons, jurymen, etc., fitted the female practitioners into their understanding of gender roles and divisions of labor. Because the city archives of Thorn burnt in 1703, and the medical files of Stadtarchiv Danzig vanished during WWII,[11] in this paper I will focus only on legal provisions for city midwives in the two aforementioned towns in the late 17th and 18th centuries.

Midwives' Oaths and Ordinances in Thorn

From the early 17th century onward in Thorn, city midwives as medical practitioners took oaths which first and foremost defined their duties. As civic officials, midwives were also regularly paid for their services. For instance, in 1601, the quarterly wage of the city's midwife (*Stadthebamme*) in Thorn was no less than 20 florins, whereas in 1700 – at least 25 florins. They were also offered free accommodation and the reimbursement of travel expenses.[12] During the late 17th and early 18th centuries, staple food items in Thorn such as bread were priced at 1.5 groschen per loaf, while a serving of meat (for an individual in a local plague hospital) cost 3 groschen.[13] Given the conversion rate of 1 florin to 30 groschen,[14] it becomes apparent that a city midwife's annual income could sustain her the purchase of approximately 400 loaves of bread or 200 servings of meat. Furthermore, the provision of lodging and the opportunity for additional earnings through private consultations enhanced the socioeconomic

---

9   Pękacka-Falkowska, Dyscyplinować i pomagać, 2013, p. 71.
10 Praetorius, Thorn, 1832, p. 391.
11 Pękacka-Falkowska, Między Gdańskiem a Toruniem, 2020, p. 118–119.
12 Pękacka-Falkowska, Dyscyplinować i pomagać, 2013, p. 75.
13 Pękacka-Falkowska, Dżuma w Toruniu, 2019, p. 253.
14 Bär, Grundbesitz, Leipzig 1911, p. 91.

status of city midwives, establishing them as relatively affluent members of the commonalty (*pospólstwo*) – a lower stratum of the local middle class.

The oldest known midwife oaths in Thorn date back to the late 17th century.[15] And indeed, before an aspiring midwife could receive a license to practice in Thorn, she had to apply to the city council, take an exam, and swear an oath of office to obey the rules of conduct laid down by the city authorities. These rules at first had little to do with the medical skills of the female practitioner, as they were mostly concerned with her character and intent. For instance, in the 17th century, a midwife in Thorn was first and foremost intended to be pious, honest and decent. She also pledged under oath to faithfully serve both wealthy and impoverished women with godliness and diligence. Her commitment centered on promptly assisting women in childbirth and ensuring their safety, irrespective of their social standing. She also stressed her obligation to handle cases of premarital conception diligently, ensuring neither the mother nor the child suffered neglect. Moreover, the midwife vowed to report such cases to city authorities impartially, demonstrating her dedication to transparency. This commitment extended to the examination of pregnant women suspected of premarital conception or abortion. Furthermore, the midwife promised to uphold these principles steadfastly, seeking divine assistance to maintain her integrity in the face of external pressures.[16]

The subsequent, early 18th century midwife's oath from Thorn referred to the character of a female practitioner as a person who should be pious, kind, loyal, hardworking, and sober. Once more, it standardized behaviors of midwives as an agent of moral control, although at the same time, unlike the former document, it also imposed on her the obligation to contact a city physician (or city surgeon) in cases of difficult deliveries. The new list of duties included, among others, the provision that the city midwife 'in serious cases [..] will consult the [...] municipal physician for advice and strive for the utmost discretion in all matters', etc.[17]

A significant change in the surveillance of midwives in Thorn took place in the second half of the 18th century. In the late 1750s or early 1760s there were

---

15 State Archives in Toruń (shortened as ATP going forward), AmT, Kat. II, I-64, f. 56, 104–106; APT, AmT, Kat. II, I-95a, f. 697–698.
16 The full German original is to be found here: Pękacka-Falkowska, Dyscyplinować i pomagać, 2013 p. 93.
17 The full German original is to be found here: Pękacka-Falkowska, Dyscyplinować i pomagać, 2013 p. 94.

plans to introduce a special *Hebammen-Ordnung* on the lines of the Russian edict *O porjadochnom uchrezhdenii babich'eva dela v pol'zu obshhestv*a (On decent establishment of midwifery for the benefit of society), developed by Panajota Condoidi (1710–1760),[18] a doctor of Greek origin and the head of the *Medicinskaja kanceljarija* (Medical Office). On the basis of the decree issued in 1754, the first Russian schools of the 'midwifery profession' were established in Moscow and St. Petersburg.

In Thorn, the person responsible for presenting councilors with the draft of the new midwifery ordinance along with the translation of the Russian decree into German was the city physician Johann Thomas Sömmerring, a graduate of the University of Leiden and a former student of Herman Boerhaave and Bernhard Siegfried Albinus. While presenting the Russian document, he discussed, i.a., the scope and character of training midwives, the language competences required from them and the scope of rights granted to female practitioners after completing their education and taking the oath. Unlike the earlier documents, the project of a new ordinance for the midwives (*Hebammen-Ordnung*) suggested that local female practitioners should be examined not only in terms of their character, but also in terms of their skill. If they lacked skills, they ought to be educated by city surgeons and municipal physicians. Consequently, it indicated the need for official representatives of the profession to acquire rudimentary anatomical and surgical knowledge, especially with regard to the use of simple surgical instruments (e.g., clippers, cups) and medications (mainly analgesics), etc. In fact, some of these recommendations were implemented by Sömmerring, who regularly invited local midwives to the autopsies of stillborns and infants who had died suddenly and unexpectedly. Such autopsies were carried out either by himself or by his collaborators – namely, city surgeons.[19]

However, it must be remembered that owing to Sömmerring, only a part of the provisions of *O porjadochnom…* by Condoidi could have been implemented in Thorn, as the decree was developed for the needs of the Russian Empire. Moreover, we do not have enough historical data to assess which articles were actually introduced in Thorn in the late 18th century.

---

18   Cf. Heine, Medicinisch-Historisches, 1851, p. 61, 130–132. Tsvelev u.a., "O poriadocznom uczriezdienii babiczyjewa diela w polzu obszczestwa" (k 250-letiju ukaza pravitielstwujuszczego sienata), "Zurnal akusherstwa i zenskich bolezniej" 2004, vol. 53, no. 1, p. 130–133.

19   Pękacka-Falkowska, Dyscyplinować i pomagać, 2013, p. 84–89.

Midwives' Oaths and Ordinances in Danzig

The oldest oath of the midwives (*Hebammeneid*) from Danzig dates back to ca. 1600. This juridical artifact provides insights into the initial responsibilities of sworn midwives in safeguarding maternal and infant health within the largest city of the Polish-Lithuanian Commonwealth. During the late 16th and early 17th centuries in Danzig, female practitioners initially demonstrated a steadfast dedication to the well-being of expectant mothers and their offspring. However, they were also mandated to report any instances of misconduct, including premarital pregnancy, abortion, infanticide, and other related offenses committed by pregnant or delivering women, and young mothers, to the municipal authorities.[20]

The aforementioned oath was changed on July 22, 1662, one year after the issuance of a new ordinance for city physicians in Danzig.[21] Under the revised oath, the local city midwives affirmed their readiness to provide immediate assistance to pregnant women, irrespective of the time or circumstances. This commitment extended to offering compassionate and impartial care to all expectant mothers, regardless of their social or economic standing. Additionally, sworn midwives pledged to safeguard against deceptive practices such as the substitution of infants and to prevent any form of harmful abortion. In cases of emergency, they vowed to collaborate closely with religious and medical authorities to ensure the safety and well-being of both mother and child.[22]

Comparable to contemporaneous developments in Thorn, where sworn midwives assumed multifaceted roles encompassing moral duties, those in Danzig were likewise entrusted with such vital responsibilities. However, it is noteworthy that despite their pivotal role in maternal and infant care, *vereidigte Hebammen* in both cities remained subject to the authority of male medical practitioners, highlighting the gendered dynamics within the domain of healthcare provision during this period.

---

20 The full German original is to be found here: Pękacka-Falkowska, Drzewiecki, Projekt porządku akuszerskiego, 2019, p. 64. https://doi.org/10.4467/12311960MN.19.015.11834 (05.06.2025).

21 Pękacka-Falkowska, Siek, Gdańska ordynacja, 2020, p. 137–168. https://doi.org/10.12775/KLIO.2020.064 (05.06.2025); Pękacka-Falkowska, Drzewiecki, Projekt porządku akuszerskiego, 2019, p. 64–65.

22 The full German original is to be found here: Pękacka-Falkowska, Drzewiecki, Projekt porządku akuszerskiego, 2019, p. 64–65.

The next legal regulations concerning the functioning of sworn midwives in Danzig were introduced at the beginning of the 18th century, when, in 1703, the *Medicinal-Ordnung E. E. Raths der Stadt Danzig* was published. This time, the emphasis was placed by the city councilors also on few issues related to the formal education and medical skills of female practitioners. The ordinance repeated the obligation for candidates to be examined by city physicians, which was first underlined in the *Ordnung der Physicorum Ordinariorum bey der Stadt Dantzigk* in the 1660s. It also ordered midwives to cooperate with city physicians, city surgeons, and apothecaries and to follow their recommendations and guidelines. They were also to abandon female individual pharmaceutical treatments, etc.[23]

Only 70 years later, on March 2, 1781, the city council of Danzig issued a stand-alone *Hebammen-Ordnung* and, for the first time in history, appointed a local *Hebammenmeister*, that is, a physician employed and paid by the municipal authorities to preside over obstetric matters. And indeed, in the late 1770s, on the rising tide of medical enlightenment and numerous changes related to the tasks and goals of the so-called medical police, the Danzig Society of Natural Sciences (*Naturforschende Gesellschaft in Danzig*, NFG), with its director Christian Sendel, made the first attempts to reform local midwifery. In 1779, the city authorities asked the members of the NFG to draft the project of a new ordinance. Due to financial constraints, lengthy discussions on the possible content of the document, and other inconveniences, the draft of the *Hebammen-Ordnung* was presented almost two years later, in March 1781. It pointed to the urgent need to combine the teaching of theory and practice while gaining of hands-on experience. Moreover, these postulates were to be carried out thanks to the establishment of a *Hebammenmeister* office, i.e., a teacher and supervisor for aspiring midwives.

The city authorities accepted the proposed document, and thus the NFG was granted the exclusive right to nominate three candidates for the newly established *Hebammenmeister* office. The physician appointed to this position was

---

23   Ibid. p. 62–63.

to receive an annual salary of 400 thalers,[24] 3/4 of which was to be transferred to him by the city treasury (*Kammerei*), while 1/4 by the NFG.[25]

The basic duties of a *Hebammenmeister* were grouped into five categories: training midwives and candidates for the profession (§ 1), assistance in difficult births (§ 2), expertise in the case of stillbirths or the death of a newborn (§ 3), in the winter months, conducting dissections as a part of training process (§ 4), and admitting young women to study (§ 5). Physicians appointed to the office of *Hebammenmeister* were obliged to fulfill these obligations by taking an oath.[26] Thus, the authors of the first volume of the journal 'Archiv der medizinischen Polizei und der gemeinnützigen Arzneikunde', were very enthusiastic about the newly created office.[27]

However, the situation in Danzig was not as optimistic as it might seem. According to the testimony of Franz Christian Brunatti, who returned to his native city in 1796 after an almost seven-year *peregrinatio medica*, the first *Hebammenmeister* in Danzig, doctor Martin Jacob Kubas, was disappointed with his office. The reason was that he was obliged to limit his previous professional activities as a physician in order to focus on his new teaching duties. He was also required to regularly visit poor pregnant women together with his mentees; and was forbidden to accept remuneration from pregnant women and their family members for help in childbirth.

Thus, due to his insufficient salary, he was very reluctant to carry out his duties. Moreover, he also complained about other working conditions. For instance, female students of the *Hebammenmeister* were to be instructed by him in physical examinations and the anatomy of a woman's body. In the latter case, the teacher was obliged to conduct anatomical demonstrations on the bodies of deceased women who had died in local hospitals. However, the basic training cycle lasted six months, and sometimes no women in labor died within

---

24 In the 1780s, within the city of Danzig, the exchange rate was approximately one thaler to 144 groschen. During this period, common goods held varying price points: for instance, one pound of butter typically cost 15 groschen, while one pound of coffee was priced at 39 groschen. Additionally, a set of 15 eggs could be acquired for 10 groschen. Further details on prices during this time period can be found in Furtak, Ceny w Gdańsku, 1937.
25 Brunatti, Entbindungs-Lehranstalt 1904, Bd. XI, H. 1/2, p. 41–47.
26 Pękacka-Falkowska, Drzewiecki, Projekt porządku akuszerskiego, 2019, p. 57–59.
27 Kurze Nachrichten (Nr. 23), Arzneikunde, 1783, Bd. 1, p. 351–352.

that time-span.[28] Moreover, in 1781 no birthing center (*Gebäranstalt*) was established in Danzig in which the *Hebammenmeister* could have conducted practical classes.[29]

The list of the duties of local midwives was more extensive than the list of the *Hebammenmeister's* obligations. It was also published in March 1781. Firstly, the new Danzig instruction tried to regulate the behaviour of female practitioners in relation to the *Hebammenmeister* (§ 1), themselves (§ 2), other midwives (§ 3), and pregnant women, women in labor and those in the postpartum period (§ 4, 5). In addition to the numerous procedures for dealing with difficult births (§ 5), the new document also defined the procedures in cases of the apparent death of a newborn and in situations when surgical intervention was needed (§ 6, 8). It defined the ways for using internal medications (§ 7), caring for the mother and her newborn (§ 9), dealing with the births of so-called monsters, i.e. fetuses with developmental defects (§ 10), participating in classes given by a midwife master (§ 11, 12, 13), and providing emergency baptism to newborns who were in danger of death (§ 14). Another group of provisions referred to the occupational structure of Danzig midwifery, distinguishing between sworn midwives (*geschworene Hebamme*) and the so-called ordinary midwives (*gewöhnliche Hebamme*) (§ 15); they also defined the duties of sworn midwives to their mentees (*Lehrtochter*) (§ 16) and vice-versa (§ 17). The instruction also discussed the entrance requirements for *Lehrtochter*, including the necessity to pay a fee (§ 18), the organization of training, examinations, and obtaining qualification attests (§ 19, 20, 21, 22), the obligation to take an oath after passing the final examination (§ 23), and the need for further education for the next three years after obtaining the patent (§ 24).[30]

Conclusion

The examination of midwifery in Polish Prussia reveals a complex evolutionary trajectory influenced by intricate interactions among social and medical dy-

---

28 Yet, women who wanted to become sworn city midwives, had to participate in Hebammen-Meister classes for three years or longer. During that time, they were obliged to take exams every year. However, every four years after taking up the city midwife office, they had to repeat a six-month course several times.
29 Pękacka-Falkowska, Drzewiecki, Projekt porządku akuszerskiego, 2019, p. 54–55.
30 Pękacka-Falkowska, Drzewiecki, Projekt porządku akuszerskiego, 2019, p. 59–62.

namics. A comparative analysis of Thorn and Danzig highlights the emergence of licensed midwives and the gradual establishment of regulatory legislative frameworks governing their practice in both cities.

Notably, Thorn and Danzig demonstrated rapid advancements towards formalizing the midwifery profession during the early modern period, as evidenced by the formulation of specific regulations by their respective city councils to supervise maternal healthcare. In contrast, other major cities within the Polish-Lithuanian Commonwealth experienced delays, with similar processes unfolding only at the turn of the 18th and 19th centuries.

An important aspect illuminated by the cases of Thorn and Danzig is also the gendered framework underpinning healthcare provision, wherein midwives, despite their roles in maternal and infant care, remained subordinate to male medical practitioners, namely municipal physicians and surgeons, reflecting entrenched societal norms regarding gender roles and labor divisions. This phenomenon, however, was not unique to Polish Prussia but characteristic of broader trends across early modern Western Europe.

Furthermore, the evolving regulatory legislative frameworks in Thorn and Danzig mirror broader trends in medicalisation, moralization, and population policy. The transition from character-based midwifery oaths to ordinances prioritizing medical skills, proficiency, and collaboration with municipal physicians and city surgeons indicated a shift towards the professionalization and standardization of midwifery practice. Initiatives such as the establishment of the *Hebammenmeister* office in 18th-century Danzig reflect early efforts by local authorities towards transforming midwifery into profession, albeit accompanied by numerous infrastructural and economic challenges.

It is important to recognize that city midwives in Polish Prussia throughout the entire analysed period also assumed the role of monitoring and enforcing reproductive norms among local women, effectively serving as 'moral officers.' This multifaceted responsibility demonstrates the complex societal expectations placed upon midwives beyond their medical duties as moral experts.

In conclusion, the examination of midwifery in Thorn and Danzig offers numerous insights into the interplay of gender, medicalized reproduction policy, and society within German-speaking regions of the Polish-Lithuanian Commonwealth. Nonetheless, further exploration of archival records and historical documents is essential to fully comprehend the complexities of midwifery practice and its impact on (female) healthcare. Thus, of particular interest are the legal dimensions of midwifery within Danzig, Thorn, and other cities of Polish Prussia, and the broader Polish-Lithuanian Commonwealth, including midwives' roles as expert witnesses in legal proceedings related to rape, premarital conception, infanticide, and abortion. Additionally, analyzing cases of alleged medical negligence and errors by midwives, as well as discussion their actual position within the contemporary medical hierarchy, warrants scholarly attention.

Bibliography

Kurze Nachrichten (Nr. 23), in: Archiv der medizinischen Polizei und der gemeinnützigen Arzneikunde, 1 (1783), p. 351f.

ADAMSKI, J, Rozwój położnictwa w Polsce, in: Pielęgniarka i Położna, 29 (1961), p. 14–32.

BÄR, Max, Der Adel und der adlige Grundbesitz in Polnisch-Preussen zur Zeit der preussischen Besitzergreifung. nach Ausz. aus d. Vasallenlisten und Grundbüchern, Leipzig, Hirzel, Mitteilungen der Preussischen Archivverwaltung, 1911.

BRUNATTI, Franz Christian, Die Entbindungs-Lehranstalt von Westpreußen bis zum Jahre 1825. nach seiner Original-Handschrift veröffentlicht von Dr. Rudolf Köstlin, Danzig, s.n., Aus. Schriften d. Naturforsch. Gesellschaft in Danzig, N.F., 11,1/2, 1904.

BRZEZIŃSKI, Tadeusz (ed.), Historia medycyny, Warszawa, Państ. Zakład Wydawnictw Lekarskich, 1988.

CWIELEW, J.W., ABASZYN, W.G. & KALCZENKO, A.P., O poriadocznom uczriezdienii babicyjewa diela w polzu obszczestwa" (k 250-letiju ukaza pravitielstwujuszczego sienata), in: Žurnal" akušerstva i ženskih" boleznej. Журналъ акушерства и женскихъ болезней, 53 (2004), p. 130–133.

FLACHENECKER, Helmut/KOPIŃSKI, Krzysztof (eds.), Editionswissenschaftliches Kolloquium 2021. Fortführung alter Editionsvorhaben im neuen Gewande, Toruń, TNT, Publikationen des Deutsch-Polnischen Gesprächskreises für Quellenedition = Publikacje Niemiecko-Polskiej Grupy Dyskusyjnej do Spraw Edycji Źródeł, 2022.

FURTAK, Tadeusz, Ceny w Gdańsku w latach 1701–1815, Lwów; Warszawa, Skład główny Kasa im. Rektora J. Mianowskiego-Instytut Popierania Polskiej Twórczości Naukowej, (Badania z Dziejów Społecznych i Gospodarczych, Bd. 22), 1935.

HARATYK, Anna/ZAÂČIVS'KA, Nadìâ Mihajlìvna (eds.), Rozwój polskiej i ukraińskiej teorii i praktyki pedagogicznej na przestrzeni XIX-XXI wieku. T. 7. Kultura w edukacji, Wrocław, Oficyna Wydawnicza Atut – Wrocławskie Oświatowe, 2017.

HEINE, Maximilian, Medicinisch-historisches aus Russland, St. Petersburg, Eggers, 1851.

JUSTYNIARSKA-CHOJAK, Katarzyna, Troska o zdrowie kobiet w polskich zielnikach z XVI wieku, in: Per mulierem. kobieta w dawnej Polsce – w średniowieczu i w dobie staropolskiej, ed. Katarzyna JUSTYNIARSKA-CHOJAK/ Sylwia KONARSKA-ZIMNICKA, Warszawa, DiG, 2012, p. 407–420.

JUSTYNIARSKA-CHOJAK, Katarzyna/KONARSKA-ZIMNICKA, Sylwia (eds.), Per mulierem. kobieta w dawnej Polsce – w średniowieczu i w dobie staropolskiej, Warszawa, DiG, 2012.

KARPIŃSKI, Andrzej (ed.), Wśród córek Eskulapa. szkice z dziejów medycyny i higieny w Rzeczypospolitej XVI-XVIII wieku, Warszawa, DiG. Instytut Historyczny UW, Fasciculi Historici Novi, 2009.

KOCELA, Weronika, Trudna sztuka babienia. kultura medyczna Polski II połowy XVIII wieku, Warszawa, DiG, 2020.

KOLKA, Wojciech Piotr, Klinika Położniczo-Ginekologiczna Uniwersytetu Jagiellońskiego w latach 1918–1939, Kraków, Zakład Historii Medycyny UJ, (Rozprawy z Historii Medycyny i Filozofii Medycyny = Dissertationes Medico-Historicae, Bd. 10), 2001.

KUKLO, Cezary, Badania nad historią kobiet w Polsce XVI–XVIII wieku w latach 2011–2020. Niezmienna atrakcyjność, ale czy nowe pytania?, in: Acta Universitatis Lodziensis. Folia Historica, (2020), p. 13–57.

KUROWSKA, Hanna, Akuszerka na ziemiach polskich w świetle przepisów oraz literatury medycznej z końca XVIII i pierwszej połowy XIX wieku, in: Studia Zachodnie, 17 (2015), p. 219–238.

KUŹMA-MARKOWSKA, Sylwia (1979-), Walka z "babkami" o zdrowie kobiet. medykalizacja przerywania ciąży w Polsce w latach pięćdziesiątych i sześćdziesiątych XX wieku, in: Polska 1944/45-1989. studia i materiały, 15 (2017), p. 189–215.

ŁAPIŃSKI, Z, Stan położnictwa przed powstaniem szkół położniczych, in: Polski Tygodnik Lekarski, 31 (1976), p. 1697–1599.

MATANIAK, Mateusz, Akuszerki rządowe Wolnego Miasta Krakowa (1815–1846) Z dziejów publicznej służby zdrowia na ziemiach polskich w XIX wieku, in: Krakowskie Studia Małopolskie, 22 (2017), p. 162–181.

MATUSZEWSKA, Eleonora, Zarys historii zawodu położnej, Warszawa, REA, 2012.

PĘKACKA-FALKOWSKA, Katarzyna, Das Danziger Arzneibuch aus dem Jahre 1665, in: Editionswissenschaftliches Kolloquium 2021, 2022, p. 187–189, 202–204.

PĘKACKA-FALKOWSKA, Katarzyna, Między Gdańskiem a Toruniem. Georg Seger i anatomia, in: Historia to (nie) fraszka. studia ofiarowane profesorowi Krzysztofowi Mikulskiemu z okazji 60 rocznicy urodzin, ed. Michał TARGOWSKI/Agnieszka ZIELIŃSKA-NOWICKA, Wydanie pierwsze., Toruń, Naukowe Uniwersytetu Mikołaja Kopernika, 2020, p. 117–224.

PĘKACKA-FALKOWSKA, Katarzyna, Dżuma w Toruniu w trakcie III wojny północnej, Lublin, Towarzystwo Naukowe Katolickiego Uniwersytetu Lubelskiego Jana Pawła II, (Źródła i Monografie / Towarzystwo Naukowe Katolickiego Uniwersytetu Lubelskiego Jana Pawła II, Bd. 485), 2019.

PĘKACKA-FALKOWSKA, Katarzyna, Dyscyplinować i pomagać – toruńskie akuszerki miejskie w XVIII w. (kilka uwag na marginesie przysiąg i porządków akuszerskich), in: Medycyna Nowożytna, 19 (2013), p. 65–105.

PĘKACKA-FALKOWSKA, Katarzyna & DRZEWIECKI, Bartosz, Projekt porządku akuszerskiego Towarzystwa Przyrodniczego w Gdańsku (Naturforschende Gesellschaft) z 1781 r., in: Medycyna Nowożytna, 25 (2019), p. 47–66.

PĘKACKA-FALKOWSKA, Katarzyna & SIEK, Bartłomiej, Gdańska ordynacja dla fizyków miejskich z 1661 roku, in: Klio – Czasopismo Poświęcone Dziejom Polski i Powszechnym, 56 (2020), p. 137–168.

PRAETORIUS, Karl Gotthelf, WERNICKE, Julius Emil & LOHDE, Wilhelm Theodor, Topographisch-historisch-statistische Beschreibung der Stadt Thorn und ihres Gebietes. die Vorzeit und Gegenwart umfassend. H. 1–3, Thorn, wydawca nieznany, 1832.

STAWIAK-OSOSIŃSKA, Małgorzata, Sztuka położnicza dla kobiet. kształcenie akuszerek na ziemiach polskich w dobie niewoli narodowej (1773–1914), Warszawa, DiG, 2019.

STAWIAK-OSOSIŃSKA, Małgorzata, Początki kształcenia akuszerek we Lwowie (1773–1805), in: Rozwój polskiej i ukraińskiej teorii i praktyki pedagogicznej na przestrzeni XIX-XXI wieku. T. 7. Kultura w edukacji, ed. Anna HARATYK/Nadìâ Mihajlìvna ZAÂČÌVS'KA, Wrocław, Oficyna Wydawnicza Atut – Wrocławskie Oświatowe, 2017, p. 157–171.

SZARSZEWSKI, Adam, Proces medykalizacji szpitali gdańskich. aspekty socjalne, prawne i ekonomiczne (1755–1874) = Medicalization process of the Gdansk hospitals. social, legal and economic aspects (1755–1874). rozprawa habilitacyjna, Gdańsk, Akademia Medyczna, Annales Academiae Medicae Gedanensis, 2007.

TARGOWSKI, Michał/ZIELIŃSKA-NOWICKA, Agnieszka (eds.), Historia to (nie) fraszka. studia ofiarowane profesorowi Krzysztofowi Mikulskiemu z okazji 60 rocznicy urodzin, Wydanie pierwsze., Toruń, Naukowe Uniwersytetu Mikołaja Kopernika, 2020.

TRZPIOT, Wiktoria, Stan błogosławiony czy "nieznośne męki"? Kobieta staropolska w okresie ciąży i połogu w świetle memuarów, in: Per mulierem. kobieta w dawnej Polsce – w średniowieczu i w dobie staropolskiej, ed. Katarzyna JUSTYNIARSKA-CHOJAK/Sylwia KONARSKA-ZIMNICKA, Warszawa, DiG, 2012, p. 433–449.

WASZYŃSKI, Edmund, Historia położnictwa i ginekologii w Polsce, Wrocław, Volumed, Historia Medycyny, 2000.

WĘGLORZ, Jakub, Staropolskie poradniki medyczne o zdrowiu i chorobach kobiet, in: Per mulierem. kobieta w dawnej Polsce – w średniowieczu i w dobie staropolskiej, ed. Katarzyna JUSTYNIARSKA-CHOJAK/Sylwia KONARSKA-ZIMNICKA, Warszawa, DiG, 2012, p. 412–432.

WRZEŚNIEWSKA, Marzena & BĄK, Beata, Historia zawodu położnej i kształtowanie się opieki okołoporodowej na świecie i w Polsce = History of midwifery profession and the development perinatal care in domestic and international perspective, in: Studia Medyczne, (2012), p. 89–99.

WYŻSZA SZKOŁA PEDAGOGICZNA IM. TADEUSZA KOTARBIŃSKIEGO (ZIELONA GÓRA)/UNIWERSYTET ZIELONOGÓRSKI (eds.), Studia Zachodnie, Zielona Góra, Wydaw. WSP, 1992.

ZABOROWSKA, Bożena, Pomoc przy porodach w Rzeczypospolitej w epoce nowożytnej w świetle zielników i poradników medycznych, in: Wśród córek Eskulapa. Szkice z dziejów medycyny i higieny w Rzeczpospolitej XVI–XVIII wieku, ed. Andrzej KARPIŃSKI, Warszawa, 2009, p. 279–312.

From Sex-Driven Maids to Population Regulation to the Creation of the Housewife
Reproductive Struggles in Saxony and the Habsburg Empire in the 18th Century

Tim Rütten[1]

Abstract *This article explores 18th-century discourses surrounding reproduction in Saxony and the Habsburg Empire, with an emphasis on maids as pivotal figures in domestic and biological reproduction. It examines how cultural, religious, and socio-economic factors shaped perceptions of sexuality, population policy, and gender roles. The evolving image of the housewife as the ideal representation of reproductive order, reflecting a broader shift towards the regulation of female labor and moral.*

Introduction

This article is devoted to urban discourses on reproduction in what is today eastern and southern Germany, as well as the former Habsburg territories, during the 18th century. Reproduction is broadly understood as sexual reproduction and/or the work of reproducing. Maids occupied a key position in domestic reproduction; they cooked, cleaned, and took care of children. Additionally, maids could reproduce themselves, which often created tension among their employers and society at large.

There is a large body of research available on servants and their behavior, especially in the European context,[2] including a preponderance of studies con-

---

1 Universität der Bundeswehr Munich.
2 For an overview, See: Sarti, Historians, 2014, p. 279–314.

centrating on the turn of the 20th century.[3] Work on the early modern period, which proceeds discourse-analytically, serves to fill a gap in research.[4] Until now, the social-demographic perspective has been the predominant approach to tackling questions of reproduction.[5]

To begin, I will narrow down the discourse and attempt to anchor it in the local socio-historical context. Following this, I will explore the connection between biological, cultural, and social ideas about reproduction. I argue that changing socioeconomic factors led to a convergence of Catholic and Protestant discourse, making the reproductive behavior of poor people a problem. The article explores debates surrounding reproduction in the early modern era, revealing that the discourse was dominated by cultural and religious factors rather than medical considerations. As the article notes, in the 18th century, conversations about sexuality and reproduction within families were still a matter of ongoing negotiation. Yet, the delayed debates in Leipzig from 1680 to 1720 and in Vienna from 1780 to 1810 provide evidence that biological perspectives on reproduction and motherhood were slowly gaining wider acceptance, gradually migrating eastward and ultimately supplanting religious interpretations.

Querelle des Servantes

The discourse on servants extends into the Middle Ages, using ancient and biblical evidence[6] that can easily be categorized as Protestant, secular, and Catholic. The dominant powers in Saxony, Protestant and secular, sought to diminish the honor of maids, and as a result, expand the lordly power. Catholicism, which dominated the Habsburg Empire, maintained a counter-discourse until the 1750s. Catholic priests praised servants and valued their labor.[7]

---

3   This is still true for the opulent: Pasleau, Schopp, Sarti, Modelization, Liège 2005. Exceptions: Frühsorge u.a. (eds.), Gesinde, 1995.
4   Cf. Whittle (ed.), Servants, 2017. But the related topic of the house has received attention: Eibach, Lanzinger (eds.), Routledge History, 2020.
5   Cf. Fauve-Chamoux, Illegitimacy, 2011, p. 8–44.
6   Cf. Glaser, Gesind=Teufel, 1564.
7   The proof can be provided indirectly by considering the work of maidservants in Catholic texts as equivalent to spiritual or other work. Cf. Dienstbotten=Schul, 1755, p.

Over the course of the 17th century, this two-gender discourse[8] evolved first from its Protestant and secular origins into a mono-gendered form that focused on maids.[9] This change is due to socio-economic movements, such as migration and cultural changes, and discussions about female education.[10] It is connected to the *Querelle des Femmes*,[11] and is in this text referred to as the *Querelle des servantes*. The discourse positioned maids as disruptive factors. As working women, they stood in contrast to the burgeoning image of a homely mother. As liminal figures, maids helped loosen the constraints of gender norms of the period.[12]

Saxony and Leipzig

The Thirty Years War gave way to a challenging social situation: many areas remained underpopulated,[13] city treasuries were empty,[14] wages were low, and prices for grain were high.[15] The climate was poor. Nevertheless an unrestrained population growth continued until the 1750s.[16] The population of people living in poverty grew in many cities up to 50% by 1700.[17] Leipzig itself went from 24,000 to 30,000 inhabitants and was a magnet for the poor.[18] Many of these migrants were rural women trying to find employment, which led to further imbalance in the marriage market with a general surplus of women in cities.[19] The poor social situation impacted the servant system. Private individuals, as well as those close to the state, made attempts since the 1650s to solve problems by renewing the servants' orders.[20] Uncontrolled

---

23–30. Alternatively, one can prefer the industrious maid to the vain prince: Rauscher, Oel, 1689, p. 335.
8   Not gendered: Glaser, Gesind=Teufel, 1564.
9   e.g.: Marforius, Kurtze Beschreibung.
10 Kreis-Schinck, Frauenbildung, 1996, p. 15–27.
11 Bock, Zimmermann, Querelle, 1997, p. 9–38.
12 Cf. Rütten, Geschäft, 2022, p. 45–47.
13 Cf. Pfister, Bevölkerungsgeschichte 2007, p. 14–17.
14 Hoffmann-Rehnitz, Unwahrscheinlichkeit, 2016, p. 169–208.
15 Cf. Waschinski, Währung, 1952, p. 154.
16 Schilling/Ehrenpreis, Stadt, 2015, p. 11–16.
17 Mörke, Social structure, 1996, p. 150.
18 Kervorkian, Poor, 2000, p. 164f.
19 Fauve-Chamoux, surplus urbain, 1998, p. 359–377.
20 Cf. Wuttke, Gesindeordnungen, 1893, p. 61–146.

marriage and migration were seen as interwoven phenomena. The orders tried to curb uncontrolled migration by making registration compulsory.[21] In Leipzig, a center for book printing, a number of writings were produced,[22] which accompanied the decision-making process and hoped to influence the reproductive behavior.

The Reproductive Theater

Discussions about marriages, maids, and their offspring illustrate quite vividly the intertwining of biological, cultural, and social ideas about reproduction. Around 1700, the religious superstructure was quite evident and shaped the focus in each area.

By 1700, a maid's eligibility for marriage was a primary topic of discussion.[23] The question was a cultural-social one: should (elite) men marry maids?[24] Protestant and secular authors mostly argued to be careful due to the lack of marriage barriers between different ranks.[25] Texts berating maids therefore accused them of giving birth to "whore children" and "foundlings."[26] They claimed maids were procreating illegitimately, which was, of course, a sin. In a novel written in the style of Grimmelshausen's Courasche, set around the fictional Salinde, which is likely meant to reference Leipzig,[27] modern viewers get a glimpse of the problem. A young, fallen, formerly bourgeois, orphan daughter goes on a sexually adventurous journey. She becomes a prostitute and gives birth to an illegitimate child. The child dies. Regaining freedom through the child's death, she marries an innkeeper, which blurs her past sins through this marriage.[28] The fear was that with high infant mortality and mobility, a birth could easily be covered up. Since marriage functioned as

---

21 E.g. Neue Gesinde=Ordnung, 1735, Tit I § 4 u. 5 u. 6.
22 Contemporaries around 1700 and at the beginning of the 19th century perceived these writings as belonging to Leipzig. Cf. Zelander, Sünden=Blöße, p. 4f.
23 Cf. Praetorius, Dulc-Amarus, 1664.
24 For a balancing approach that argues for controlled marriage among the poor, See: Marperger, Verheyrathung, 1717, p. 11f.
25 Against any barriers and in favor of a marriage with a maid, See Praetorius, Dulc-Amarus, p. 25–46. Against any marriage with maids, see Mägde=Verfechter, B 1 [b].
26 Mägde=Verfechter, A 8 [a].
27 Cf. Hatfield, Picaras 1932, p. 515.
28 Celibilicribrifacio, Jungfer Robinsone, 1724.

a change of status, sinful behavior could be concealed. What is important: the culprits of all problems in the 1700s were the maids; in the secular Querelle des servants, the perspective that masters are complicit is mostly ignored.

But the authors did not deny the physical truth. Maids are made out of "Eve=flesh (Eva=Fleisch)".[29] Maids were permitted to feel sexual desire. But the permissibility of lust had to be regulated. The reference to flesh theologizes sexual appetite. In addition, the "horny lust of youth",[30] symbolizing the inadmissibility of desire for maids, was introduced as an idea. Maids were accused of excessive promiscuity in an effort to discredit them and ruin their reputations. The discussions were related to the discourse of state policy and population policies.[31] The attempts at arranged marriages with servants and maids were an effort to control population growth.

To this end, writings attempted to regulate the entire theater of reproduction, e.g., by establishing hygienic standards for service. The work of maids was increasingly considered dirty. This dirt was inscribed into their own flesh, so that they became dirty.[32] Maids and their relatives were considered to be hereditarily burdened.[33] In a letter from 1810 to Paul Usteri, the writer Therese Huber bemoaned the Silesian philosopher Johann Wilhelm Ritter. Ritter had died in a financially depressing situation. Huber links the situation to Ritter's marriage with his maid, and to his children: "several reasons gave him crippled wretched children who languished in obscurity."[34] Huber is addressing an old cliché about the inherited nature of maids.[35] Furthermore, the discourse addressed the changed status of prostitutes since the Reformation, as prostitution was no longer tolerated.[36] Poor, young, rural, and migrant maids were considered a recruitment pool for prostitution.[37] The network of wet-nurses was also set to disappear because of changing attitudes and laws on prostitution. Many children had established "venereal lust" because they absorbed "their fornicating wet nurses' licentious ways and lecherous

---

29   Mägde=Lob, 1688, p. 66.
30   Marperger, Verheyrathung, 1717, p. 23f.
31   Cf. Rauscher, Impopulation, 2016, p. 135 – 162. Federici, Caliban, 2004, p. 85–91.
32   Cf. Rütten, Geschäft, 2022.
33   Marforius, Beschreibung, p. 4–7.
34   Huber, Usteri 2001, p. 56.
35   Cf. Schupp, Land=Plage, 1704, p. 71.
36   Cf. van de Pol, Bürger, 2006.
37   Cf. Marforius, Beschreibung, passim.

natures".³⁸ The goal, as texts emphasize, was to get rid of prostitution/wet-nurses/maids to strengthen the natural bond between mother and child.³⁹

Vienna and the Catholic

In Catholic areas, a different reproductive regime was in place. Lifelong celibacy was desirable. Marriage discussions concerning maids did not take place around 1700. The predominant attitude of the time was one that encouraged creating many children, accepted their mortality, and was focused on the afterlife. Catholic areas had smaller populations despite higher reproduction numbers. Active population policy was foreign in Catholic areas until the middle of the 18th century. Population shortages, however, usually also meant a shortage of servants. Thus, Catholics were not interested in strict regulation in the reproductive sphere,⁴⁰ they relied on the more egalitarian community of Christians.⁴¹

Even without a political stance on reproduction, Vienna grew from 100,000 to 250,000 inhabitants during the 18th century. The increase was primarily due to immigration.⁴² The steady influx of people led to the availability of a large reserve of unskilled women.⁴³ The rate of illegitimate births was high.⁴⁴ The situation in Vienna reflected that in Leipzig between 1700 and 1720. Comprehensive legislation on servants did begin with Maria Theresa. Joseph II tried unsuccessfully to reform the servant system or to get rid of it.⁴⁵

Of Sameness and Differences

The lack of hierarchy of labor along gender lines around 1700 is striking. The Viennese Augustinian monk and preacher Abraham a Sancta Clara criticized

---

38   Cf. Zelander, Sünden=Blöße, p. 35. Similarly: Marforius, Beschreibung, p. 20f.
39   Johann Georg Joerdensen, Ammen=Miethe 1709, p. 24. Marforius, Beschreibung, p. 20.
40   Cf. Hersche, Verschwendung, 2006, p. 215–219.
41   Cf. Rauscher, Oel, 1689, p. 232.
42   Cf. Weigl, Bevölkerungswachstum, 2003, p. 110, 120, 122–124.
43   Weigl, Bevölkerungswachstum, 2003, p. 171f.
44   Weigl, Bevölkerungswachstum, 2003, p. 118.
45   Cf. Richter, Rütten, Dienst, 2021, p. 290–293.

Catholic thinking on both masters and servants around 1700.[46] This can be interpreted as a less capitalist reproductive regime. It was not yet in catholic thinking en vogue to conceptualize domestic labor as exclusively female. Nevertheless, reproductive labor tended to be feminine.[47]

When, around 1780, after a relaxation of censorship, a debate about chambermaids in Vienna gained momentum,[48] authors accessed a broad body of knowledge. First, they mono-gendered the discourse. As was the case sixty years earlier, there are arguments about responsibility in child-rearing matters.[49] At the core of the debate were male desire and female attraction.[50] The conflict now comes into a sharper focus on bourgeois gender norms.[51] Now, men need to be criticized for their desire.[52] Maids are no longer portrayed as unbridled whores; what persists is a mild suspicion of prostitution and a warning against a marriage,[53] just as the vindication of maids focuses on the natural drive to love.[54]

In Vienna, too, concerns to regulate the servants' system were shared by private individuals and the government.[55] Propositions sought to regulate the servant system via police agencies.[56] Problems were seen in migration from "the country folk" and "foreigners", and unemployment.[57] For the registry clerk Matthäus Zach, it was the combination of gender, rural origin,[58] unemployment,[59] and the lacking education of daughters by mothers.[60] Crisis-ridden times are evident in Zach's proposition when he brings the maid, household,

---

46 A S,[ancta] Clara, Etwas für Alle, 1699, p. 545–547.
47 Cf. Dienstbotten=Schul, passim.
48 For a detailed description see Gugitz, Wiener Stubenmädchenliteratur 1902/03. See also: Richter/Rütten, Exceß, p. 297f. Cf. Spennadelstich eines Stubenmädchens, 1781.
49 Rautenstrauch, Stubenmädchen, p. 14f. M., Dem Verfasser des Büchels, 1781, p. 17f.
50 [Johann Rautenstrauch, Stubenmädchen,1781, p. 6.
51 Cf. Rautenstrauch, Stubenmädchen, p. 8.
52 Cf. Rautenstrauch, Stubenmädchen, p. 12–14. Anonymous, Spennadelstich, 6. M., Dem Verfasser des Büchels, 14; 16f.
53 Rautenstrauch, Stubenmädchen, p. 11f. M., Dem Verfasser des Büchels, p. 9.
54 M., Dem Verfasser des Büchels, 10.
55 Cf. Zach, Vorschlag 1792. Giftschütz, Skizze, 1804. Cf. Zustand des Dienstbothenwesens.
56 For a rejection of this idea, See von Sonnenfels, Bemerkungen, 1810.
57 Giftschütz, Skizze; Zach, Vorschlag; Zustand des Dienstbothenwesens.
58 Zach, Vorschlag, Anleitung 1.
59 Zach, Vorschlag, Auszug.
60 Zach, Vorschlag, I. Notheissung, f. 3.

and society into conflict over the reproductive resource of food.[61] He suspects maids of stealing and therefore demands supervision by the mistress.[62] He argues, therefore, that a woman must be knowledgeable in household matters.[63] Yet talk about maids were always doubly reproductive, as not only the service-giving family was threatened, but the maid's future family would also suffer from her lack of domestic competence.[64] The community of Christians from around 1700, having given way to a competitive relationship, places families rather than the home at the center of the threats. To speak of families indicates a change in meaning. Vienna's new servants' law (1810) expanded not the power of police agencies, but of the pater familias.[65] A family is not as public as a house. Families have to be protected from external, e.g., police, influence; then the house can become a gendered reproductive refuge.

Summary

The discourse on maids was deeply embedded in everyday life, as evidenced by letters or propositions. In times of population growth, migration, or social changes, it was necessary to discredit certain channels of reproduction. Social disorder was highly perceived as a gender disorder. Attempts in Leipzig and Vienna tried to prevent the marrying off of maids. Both wanted to control young women, slow population growth, and strengthen the housewife and the home as a place of retreat.

The discourse alluded directly to the concept of the family, based on the biological concept of reproduction. When maids were portrayed as sexually conspicuous or writers demanded that mothers take care of their children, it prepared the idea that reproductive labor was the job of the biological mother, not a maid. In the 18th century, the process was not completed – almost no one demanded that service should stop. Yet developments took on a decisive spin that shaped the reproductive role of the housewife to this day.

---

61 Zach, Vorschlag, II. Notheissung, f. 6. Cf. A discussion of society: ibid.: III. Notheissung, f. 9.
62 Zach Vorschlag, III. Notheissung, f. 10.
63 Zach, Vorschlag, I. Notheissung, f. 3.
64 Zach, Vorschlag, III. Notheissung, f. 8.
65 Cf. Sonnenfels, Bemerkungen, p. XIII f.; p. 64f, 80f.

Sources

Österreichisches Staatsarchiv AT-OeStA/HHSTA KA KFA 67 Wohlfeilsachen (alt 67) und diverse Projekte und Schriften (alt 68), 1791–1793, 9 Vorschlag des Mathäus Zach zu einer guten Dienstbotenordnung, 1792.02.04.

Wiener Stadt- und Landesarchiv, A1 – Hauptarchiv – Akten und Verträge: Hist.5/1804, Anton Friedrich Giftschütz, Skizze zu einer neuen Dienstbothen=Ordnung für die Stadt Wien und ihre Vorstädte.

A S.[ANCTA] CLARA, Abraham, Etwas für Alle / Das ist: Eine kurtze Beschreibung allerley Stands- Ambts- und Gewerbs-Personen / Mit beygedruckter sittlicher Lehre und Biblischen Concepten [...], Würzburg 1699.

Der Unbescheidene Mägde=Verfechter, Wird hiermit samt seinen Leutseligen Muhmen, Getreuen Ammen, Verständigen Köchinnen, Arbeitsamen Junge=Mägde, Höflichen Jungfer=Mägden Und Lehr=begierigen Kinder=Mägden, Bescheiden nach Hause gewiesen Von Marforio guten Freunde, [s.l.] [s.a.].

Mägde=Lob Oder: Der Dienst=Mägde Unschuld / In allen Unter= und Ober=Gewehren tapffer verfochten / und mit annemlichern Farben / dem neuligst hervorgeschlossenen Tractätgen entgegen gesetzet, (s.l.) 1688.

Ueber den gegenwärtigen Zustand des Dienstbothenwesens in Wien und die Mittel zur Verbesserung desselben, in: Vaterländische Blätter für den österreichischen Kaiserstaat XV (28. Juni 1808) p. 119–124, XVI (1. Juli 1808), p. 127–132; 17 (5. Juli 1808), p. 139–142.

GIFTSCHÜTZ, Anton Friedrich, Skizze zu einer neuen Dienstbothen=Ordnung Für die Stadt Wien und ihre Vorstädte. In: WStLa, A1 – Hauptarchiv – Akten und Verträge: Hist.5/1804.

GLASER, Peter, Gesind Teuffel Darin acht stück gehandelt werden / von des Gesindes vntrew / welche im nachfolgenden blat verzeichnet, Weissenfels [1564].

HUBER, Therese, An Paul Usteri in Zürich Günzburg 17. oder 18. Februar 1810 Sonnabend oder Sonntag, in: Magdalene Heuser (ed.), Therese Huber Briefe 4, 1810 – 1811, bearbeitet von Petra Wulbusch, Tübingen 2001, p. 52 – 57.

JOERDENSEN, Johann Georg, Die Sündliche Ammen=Miethe Dadurch Denen leiblichen Kindern, Die ihnen von GOtt und der Natur weißlich bereitete Nahrung entzogen [...], Leipzig 1709.

M., Theresia, Dem Verfasser des Büchels über die Stubenmädchen: Etwas auf die Nase. Von Theresia M. einem Stubenmädchen in Wien. Wien 1781.

MARFORIUS, Kurtze Beschreibung Des zum theil liederlichen Lebens und Wandels Derer anjetzo in grossen Städten sich befndenden Dienst=Mägde, Als da sind Muhmen oder Kinder=Frauen, Ammen, Köchinnen, Junge=Mägde, Jungfer= und Kinder=Mägden etc. [...], [s. l.] [s.a. (ca. 1717)].

MARPERGER, P[aul] J[acob], Wohlgemeynter Vorschlag..... Dienst=Mägde / Als einer GOtt zu EHren / dem Publico zum besten, und denen dabey interessirten Personen zu grossen Nutzen [...], Leipzig 1717.

RAUSCHER, Wolfgang, Oel und Wein Deß Mitleidigen Samaritans Für die Wunden der Sünder. Das ist Catholische / mit Christlichem Ernst / geistreicher Schärpffe / und Mildigkeit vermischte Predigen / zu Bekehrung und ewigen Heyl [...] 1, Dillingen 1689.

[RAUTENSTRAUCH, Johann], Ueber die Stubenmädchen in Wien, 4th edition Wien 1781.

SCHUPP, Johann Balthasar/Anonymus, Die rechte Land=Plage des heutigen Gesindes / Böser Knechte und Mägde / Von welchen Herren und Frauen anietzo / mehr als jemahln / unmenschlich geplaget / ja gepeinigt werden / [...], Leipzig 1704.

SERVIUS [Johann Praetorius], Dulc-Amarus Ancillariolis: Das ist / Der süß=wurtzligte und saur=ampferigte Mägde=Tröster [...], [s.l] 1664.

ZACH, Mathäus, Vorschlag des Mathäus Zach zu einer guten Dienstbotenordnung, 1792.02.04, in: At-OeSta7HHSTA KA KFA 67–9.

ZELANDER, Johann Georg: Die auffgedeckte Sünden-Blöße Der bißher nicht onen zur Verabscheuung, und nebst einem Anhang von denen wegen der Mägde ohnlängst in Leipzig heraus gekommenen Schrifften, allen Christen überhaupt zu erbaulicher Betrachtung vor Augen gelegt [s.l.] [1722].

Bibliography

BOCK, Gisela/ZIMMERMANN, Margarete, Die Querelle des Femmes in Europa, in: Querelles. Band 2. Die europäische Querelles des Femmes. Geschlechterdebatten seit dem 15. Jahrhundert, ed. Gisela BOCK/Margarete ZIMMERMANN/Monika KOPYCZINSKI, Stuttgart, J.B. Metzler, 1997, p. 9–38.

EIBACH, Joachim/LANZINGER, Margareth, The Routledge History of the Domestic Sphere in Europe 16th to 19th Century, Abingdon 2020.

FEDERICI, Silvia, Caliban and the witch. Women, the Body and Primitive Accumulation, New York [2004] 2009.

FREEDMAN, Joseph S. (ed.), Die Zeit um 1670. Eine Wende in der europäischen Geschichte und Kultur? (= Wolfenbütteler Forschungen 142), Wiesbaden 2016.

FRÜHSORGE, Gotthard/GRUENTER, Rainer/FREIFRAU WOLFF METTERNICH, Beatrix (eds.), Gesinde im 18. Jahrhundert (= Studien zum Achtzehnten Jahrhundert 12), Hamburg 1995.

GUGITZ, Gustav, Die Wiener Stubenmädchenliteratur von 1781. Ein Beitrag zur josephinischen Broschüren- und zur Dienstbotenliteratur, in: Zeitschrift für Bücherfreunde, 6,1 (1902/03), p. 137–150.

[HAKEN, Johann Christian Ludwig], Bibliothek der Robinsone. In zweckmäßigen Auszügen / 4, Berlin, Unger, 1807.

HATFIELD, Theodore M., Some German Picaras of the Eighteenth Century, in: The Journal of English and Germanic Philology, 31 (1932), p. 509–529.

HERSCHE, Peter, Muße und Verschwendung. Europäische Gesellschaft und Kultur im Barockzeitalter I & II, Freiburg 2006.

HOFFMANN-REHNITZ, Philip, Zur Unwahrscheinlichkeit der Krise in der Frühen Neuzeit. Niedergang, Krise und gesellschaftliche Selbstbeschreibung in innerstädtischen Auseinandersetzungen nach dem Dreißigjährigen Krieg am Beispiel Lübecks, in: Die Krise in der Frühen Neuzeit, ed. Rudolf SCHLÖGL/Philip R. HOFFMANN-REHNITZ/Eva WIEBEL, Göttingen, Vandenhoeck & Ruprecht, Historische Semantik, 2016, p. 169–208.

KEVORKIAN, Tanya, The Rise of the Poor, Weak, and Wicked. Poor Care, Punishment, Religion, and Patriarchy in Leipzig, 1700–1730, in: Journal of Social History, 34 (2000), p. 163–181.

KREIS-SCHINCK, Annette, Frauenbildung in der Frühen Neuzeit. Mary Astells "A Serious Proposal to the Ladies", in: Freiburger FrauenStudien 2 (1996), p. 15–27.

MÖRKE, Olaf, Social structure, in: Germany. A new social and economic history Vol. 3 Since 1800, London, Arnold, 2003, p. 134–163.

OGILVIE, Sheilagh C./OVERY, R. J., Germany. A new social and economic history Vol. 3 Since 1800, London, Arnold, 2003.

PASLEAU, Suzanne, SCHOPP, Isabellle, SARTI, Raffaella (eds.), The Modelization of Domestic Service. Proceedings of the Servant Project. Volume 5, Les Editions de l'Université de Liège, Liège, Belgium, 2005.

PFISTER, Christian, Bevölkerungsgeschichte und historische Demographie 1500–1800 (= Enzyklopädie Deutscher Geschichte 28), 2. edition, München 2007.

Rauscher, Peter, "'Impopulation' und 'Peuplierung'. Der Beginn staatlicher Bevölkerungspolitik von der Mitte des 17. bis zur Mitte des 18. Jahrhunderts. Die Habsburgermonarchie und Brandenburg-Preußen im Vergleich", in: Freedman, Joseph S. (ed.): Die Zeit um 1670: eine Wende in der europäischen Geschichte und Kultur?, 1. Aufl. Aufl., Wiesbaden: Harrassowitz Verlag in Kommission 2016 (Wolfenbütteler Forschungen, Band 142), p. 135–162.

Richter, Jesscica/Rütten, Tim: "[S]ie war männersüchtig, vergnügungssüchtig, unrein, faul ,bis zum Exceß' […]." Wandel und Kontinuität im häuslichen Dienst im 19. Jahrhundert, in: Elisabeth Loinig, Oliver Kühschelm, Willibald Rosner, Stefan Eminger (eds.), Niederösterreich im 19. Jahrhundert 2, St. Pölten 2021, p. 283–316.

Rütten, Tim, Ein schmutziges Geschäft, in: Österreichische Zeitschrift für Geschichtswissenschaften, 33 (2022), p. 35–57.

Sarti, Raffaella, Historians, Social Scientists, Servants, and Domestic Workers. Fifty Years of Research on Domestic and Care Work, in: International Review of Social History, 59 (2014), p. 279–314.

Schilling, Heinz/Ehrenpreis, Stefan, Die Stadt in der frühen Neuzeit, Berlin, Boston, De Gruyter Oldenbourg, Enzyklopädie deutscher Geschichte, 2015.

Schlögl, Rudolf/Hoffmann-Rehnitz, Philip R./Wiebel, Eva (eds.), Die Krise in der Frühen Neuzeit (= Historische Semantik 26), Göttingen, Vandenhoeck & Ruprecht, Historische Semantik, 2016.

Waschinski, Emil, Währung, Preisentwicklung und Kaufkraft des Geldes in Schleswig-Holstein von 1226 – 1864 (= Quellen und Forschungen zur Geschichte Schleswig-Holsteins 26), Neumünster 1952.

Whittle, Jane, Servants in Rural Europe, 1400–1900 (= People, Markets, Goods: Economies and Societies in History 11), Woodbridge 2017.

Wuttke, Robert, Gesindeordnungen und Gesindezwangsdienst in Sachsen bis zum Jahre 1835. Eine wirtschaftsgeschichtliche Studie, Leipzig, Duncker & Humblot, 1893.

Van de Pol, Lotte, Der Bürger und die Hure. Das sündige Gewerbe im Amsterdam der Frühen Neuzeit. Aus dem Niederländischen von Rosemarie Still (= Geschichte und Geschlechter Sonderband), Frankfurt a. M. 2006.

Is marriage so Sacred?
Extramarital Births in West Prussia circa 1900[1]

Hadrian Ciechanowski[2]

Abstract *This article deals with the issue of how extramarital reproductive behavior was connected with the official religion and culture cannons. The article is based on the data from the three civil status registry offices in West Prussia, where the mixed Polish, German, and Jewish population lived. Based on the data, author analyzed extramarital births among Catholics, protestants, and Jews at the turn of the 19th and 20th centuries.*

Introduction

The article's topic refers to two aspects addressed in the volume – family and reproduction. Although the article is decidedly practical, it corresponds to other texts in the volume, which treat the problems of reproduction and family planning more theoretically. On the example of selected registry office districts established by the Prussian authorities on the territory of West Prussia, the article will trace the reproductive behavior of the local population beyond the traditional canons, set by the framework of culture and religion, and therefore mainly concentrated inside the marriages.

West Prussia was considered an excellent area for this type of research. The partitions of Poland carried out in 1772, 1792, and 1795 led to the country's division between Austria, Prussia, and Russia. Part of the Prussian partition was West Prussia, which included the area of former Royal Prussia. This was an area with a mixed German and Polish population. Traditionally the German popu-

---

1   Funding: National Science Centre, Poland, grant number 2019/32/C/HS3/00121.
2   Faculty of History, Nicolaus Copernicus University Toruń, Poland.

lation was Protestant, while the Polish population was Catholic. There was also a tiny population of Jews, living mainly in the cities.[3]

These aspects make West Prussia a great research field because that territory was always a place where different cultures mixed and were transferred from Germany to Poland and back again. This situation creates a unique research opportunity and allows us to observe different populations, groups, and behaviours.

Prussia at the turn of the century was experiencing a strengthening of nationalism among German and Polish subjects.[4] This phenomenon was also associated with a religious revival among adherents of both Christian denominations. This was in turn connected with the significant development of grassroots groups that sought to educate and transmit the Protestant faith and morality among the German population.[5] On the other hand, the Polish Catholic population gathered around the Catholic Church, seeing it as a unifying force that carried forward Polish national traditions. Attachment to faith and deep devotion to it is traditionally emphasized in historiography.[6]

Traditionally, the Church's teaching placed great emphasis on the sexual sphere of believers; it paid particular attention to their moral conduct and marital fidelity. It seems, therefore, that strong commitment to church and religiosity should be reflected in fertility trends, especially in the small number of extramarital children. However, was this the reality? In this paper, the author will try to answer this question in part.

Therefore, adherence to religious practices and church teachings should be closely linked to reproductive behavior, which connects this article to the main title issues addressed in this volume: sexuality, family, and reproduction.

Methodology

The research presented in this article is based on archival materials from the State Archive in Toruń. To gather the appropriate research data five registry offices were drawn upon in the territory of West Prussia. As a random sample, the following archives may paint a picture of available research material:

---

3   Chwalba, Historia Polski 2000, p.435-465.
4   Kucharczyk, Prusy. 2020, p. 633. Frackowiak, Ethnizität, 2014, p. 39.
5   Clark, Powstanie, 2009, p. 367.
6   Kucharczyk, Prusy, 2020, p. 659.

- Registry Office Dusocin,[7]
- Registry Office Golub Zamek,[8]
- Registry Office Bierzgłowo,[9]
- Registry Office Książki,[10]
- Registry Office Małe Czyste.[11]

In the selected archives registers of births were analysed from the years 1875–1910. These years were not selected at random; 1875 was the first full year of civil registries in West Prussia. 1910, however, was chosen based on the availability of records on the Internet. Pandemic constraints did not allow for a search for records not available online. However, this is not the entire period of the functioning of Prussian Registry Offices on Polish territory; nonetheless, it was possible to see some fertility trends. The gathered data were compared to the information about fertility in the Polish People's Republic and the Republic of Poland to draw some conclusions.

The division into denominations was established according to the mother's religion. In the case of information about illegitimate children, there is usually no information about the father and his religion. The only specific information in this case, therefore, concerned the mother. That is why only she could be taken into account. All Protestant denominations were counted together.[12] The data gathered during the research is presented in the graphs in the article.

The Demographic Structure of the Districts

Prussian administration regularly took censuses of the people living in the kingdom. According to these censuses, it is possible to look at the demographic structure of the researched territory. To show the population of the drawn districts the censuses of 1885 and 1905 were compared. Each of the censuses makes it possible to trace the total population of a given registry office and its denominational structure.

---

7   State Archive in Toruń, fond numer 69/1125, Urząd Stanu Cywilnego Dusocin.
8   State Archive in Toruń, fond numer 69/1134, Urząd Stanu Cywilnego Golub Zamek.
9   State Archive in Toruń, fond numer 69/1105, Urząd Stanu Cywilnego Bierzgłowo.
10 State Archive in Toruń, fond number 69/1161, Urząd Stanu Cywilnego Książki.
11 State Archive in Toruń, fond number 69/1177, Urząd Stanu Cywilnego Małe Czyste.
12 Birth registers do not discriminate protestant denominations but only give the information a person is "ewangelic".

In 1885 the Registry Office Dusocin district was inhabited by 1,618 people, of whom 1,377 were Protestants, 232 were Catholics, and only nine were Jews.[13] By comparison, in 1905, the district had a population of 1,424. Of this population, 1,177 were of Protestant faith, and 247 declared themselves to be Catholics. In that census, however, no residents of the Mosaic faith were recorded.[14]

The Registry Office Golub Zamek had a population of 1,243 in 1885. Among them, there were 342 Protestants, 893 Catholics, and eight Jews.[15] In 1905, on the other hand, the district was inhabited by 1,528 people, 427 of whom were Protestant and 1,088 Catholic, as well as 13 Jews[16].

The next district analyzed is the Registry Office Bierzgłowo. In 1885 it had 1,892 inhabitants. There were 498 Protestants, 1,378 Catholics, and 16 Jews as far as religion is concerned.[17] In 1905, however, it was inhabited by 2,297 people. Among them, there were 815 Protestants, 1,477 Catholics, and five Jews.[18]

According to the 1885 census, the population of Książki was 1,939. 1,594 of them were Protestants, 177 were Christians of other kinds, 157 Catholics, and 11 Jews.[19] By contrast, in 1905 there were 2,234 people, of whom 1,872 were of the Protestant faith, 139 of other Christian faiths, and 223 of the Catholic faith.[20]

In the Registry Office Male Czyste, there were 1,612 inhabitants in 1885. Among them, there were 835 Protestants and 777 Catholics.[21] In 1905, on the other hand, there were only 1,503 people. Of these, 804 were Protestants, and 699 were Catholics.[22]

---

13   Gemeindelexikon, 1887, p. 126–135.
14   Gemeindelexikon, 1908, p. 52–55.
15   Gemeindelexikon, 1887, p. 106–111.
16   Gemeindelexikon, 1908, p. 8–15.
17   Gemeindelexikon, 1887, p. 112–119.
18   Gemeindelexikon, 1908, p. 142–145.
19   Gemeindelexikon, 1887, p. 106–109. It is not clarified in the source what "other Christians" means.
20   Gemeindelexikon, 1908, p. 8–13 It is not clarified in the source what "other Christians" means.
21   Gemeindelexikon, 1887, p. 120–127.
22   Gemeindelexikon, 1908, p. 14–19.

Summarizing the data collected, the population living in the study area is as follows:

Table 1: Population living in the study area

Registry office	Population		Protestants		Catholics		Jews	
Year	1885	1905	1885	1905	1885	1905	1885	1905
Dusocin	1618	1424	1377	1177	232	247	9	0
Golub Zamek	1243	1528	342	427	893	1088	8	13
Bierzgłowo	1892	2297	498	815	1378	1477	16	5
Książki	1939	2234	1771	2011	157	223	11	0
Małe Czyste	1612	1503	835	804	777	699	0	0
Together	8304	8986	4823	5234	3437	3734	44	18
Percent	100%	100%	58,08%	58,25%	41,39%	41,55%	0,53%	0,2%

Source: own study

The censuses show that Protestants dominated in the chosen districts, far outnumbering Catholics. Jews, on the other hand, made up only a fraction of one per cent of the total population. Generally, these proportions remained unchanged during the whole period under study. As far as particular registry offices are concerned, Protestants prevailed in Registry Office Dusocin and Registry Książki. On the other hand, there was a predominance of Catholics in Registry Offices Bierzglowo and Golub Zamek. In Registry Office Male Czyste, the population of both denominations remained in relative balance, with a slight Protestant advantage.

Fertility Rates in Registry Offices

The first point of the analysis looks at fertility rates among religious groups in each selected registry office.

Figure 1: Registry Office Dusocin fertility rates by religion

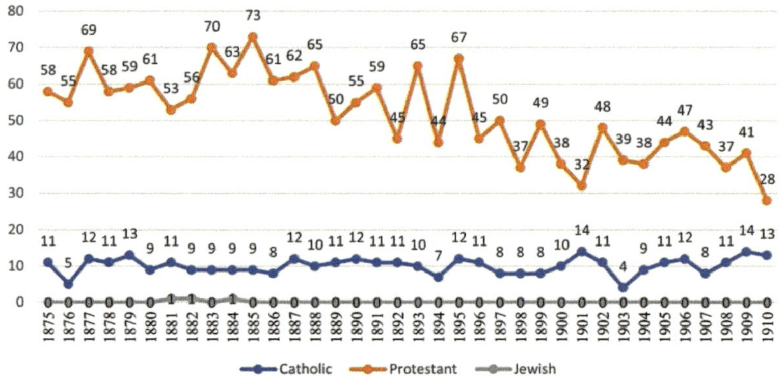

Source: own research

The gathered data shows that, during the entire period, in the Registry Office Dusocin there was a clear predominance of births to children of Protestant denominations. However, from 1896 we note a clear decreasing trend in the number of Protestant births. At the end of the period, we find a 30% drop. In contrast, the number of Catholic births was almost constant throughout the whole period. The number of Jewish births was marginal. There were only three such cases during the entire period under study, one each in 1881, 1882, and 1884.

Figure 2: Registry Office Książki fertility rates by religion

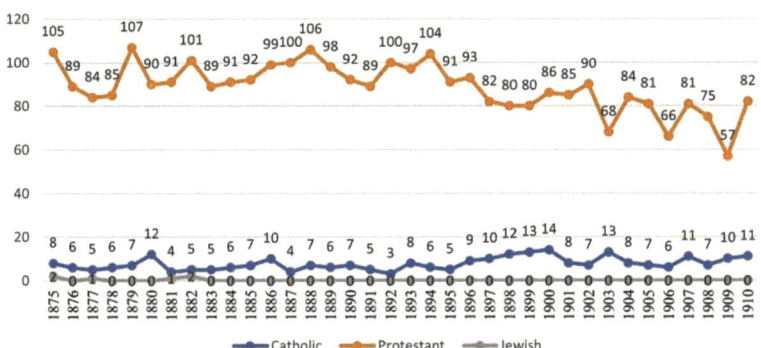

Source: own research

A similar trend can be observed in the Registry Office Książki. However, in this district, the decrease in Protestant births was not as significant as in the Registry Office Dusocin. Except for 1903, 1906, and 1909, it oscillates between 10% and 26%. Furthermore, in this registry office, the number of catholic births was constant throughout the whole period. As in the Dusocin, the number of Jewish births was minimal. There were recorded only six births – two per year in 1875 and 1882, and in 1875, one each in 1877 and 1881.

Figure 3: Registry Office Bierzgłowo fertility rates by religion

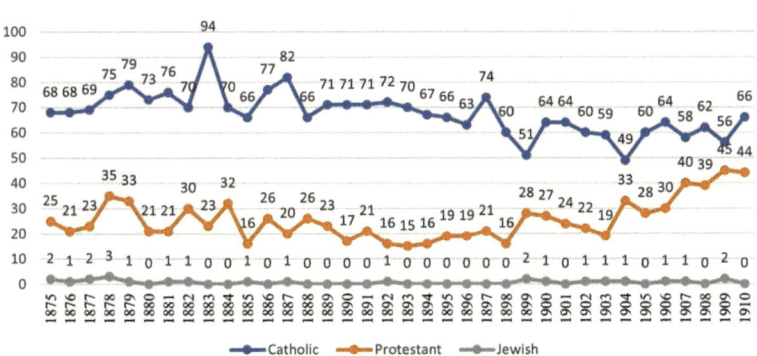

Source: own research

Unlike the previous two districts, in the Registry Office Bierzglowo, the dominant part of the population throughout the entire period under study was that of Catholics. However, the observed disproportion between denominations was smaller than the previous. The observed trend shows that from 1897 the number of Catholic births was slowly decreasing. At the same time, the number of Protestant births was growing. This growth from the beginning to the end of the research period was almost twofold. Jewish births are slightly higher for this Registry Office and were recorded more frequently, sometimes even annually. Compared to Catholics and Protestants, however, this number is still minimal.

Figure 4: Registry Office Golub Zamek fertility rates by religion

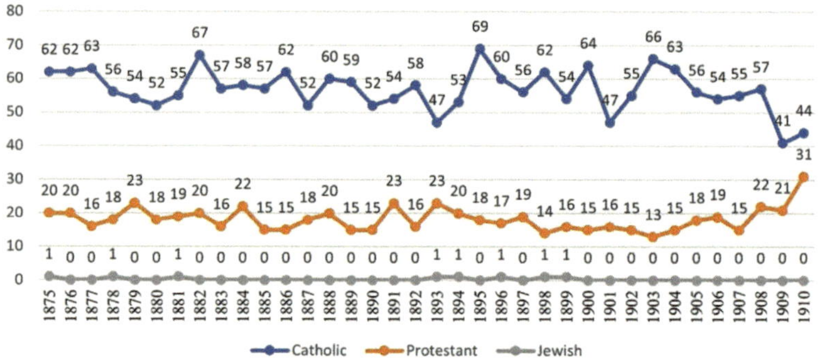

Source: own research

Catholic births also dominated in the Registry Office Golub Zamek. Their number was almost constant throughout the entire period. A sharp decline was noted only in the last two years, but the lack of data from the following years makes it impossible to say that this was a trend. Also, more or less constant was the number of Protestant births. In the whole period, Protestant births account for 30 to 50% of Catholic births. As in the previous districts, the number of births to Jewish children is minimal. They occurred only eight times during the entire period.

Figure 5: Registry Office Male Czyste fertility rates by religion

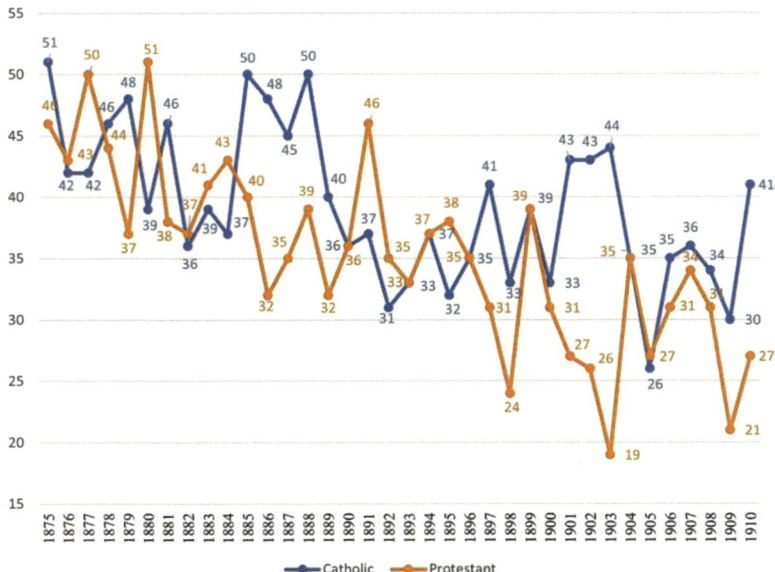

Source: own research

In the last drawn district, the situation was slightly different than in the other registry offices. The Registry Office Male Czyste was the only office studied in which were no clearly dominant denomination. However, in both Christian religions the number of births was slowly decreasing. In the whole period there were 142 more Catholic births. This is very interesting because in this registry office were living more Protestants, and the number of Catholics declined more significantly than Protestants between 1885 and 1905.

As can be seen, the number of births in all registry offices corresponds with the demographic structure of the districts under study. However, it is necessary to integrate the data for the whole territory to analyze the research problem. Chart 6 shows the total number of marital and extramarital births.

Figure 6: Total number of marital and extramarital births

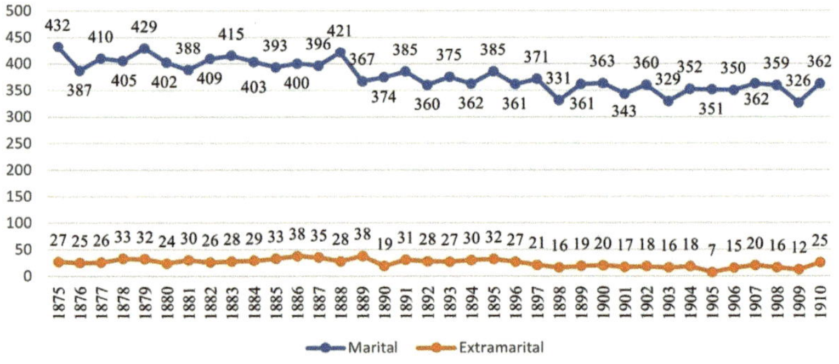

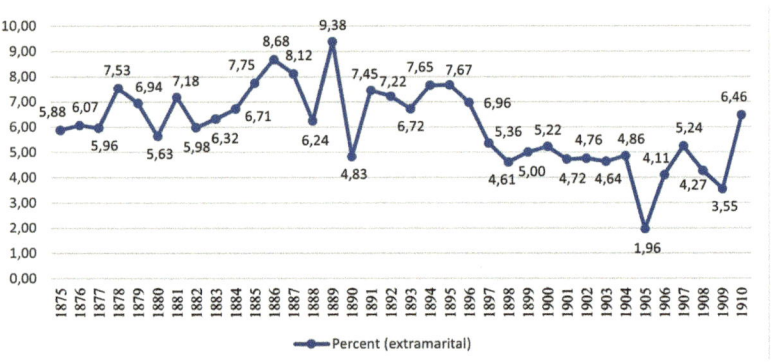

Source: own research

Figure 7: Total percent of extramarital births

Source: own research

Looking at the absolute numbers of the population between the beginning and the end of the research period, a clear downward trend in the number of births and births to legally registered unions can be observed. While in 1875, 459 children were born, of whom 432 were "legitimate", in 1910 only 387 children were born, of whom 362 were legitimate. The difference in the number of births was therefore 72, compared to 70 for legitimate births. The number of extramarital births was relatively stable, fluctuating only slightly, especially towards the end of the period under study. It seems that the decline in the total number of births may have been caused by the general depopulation of the eastern provinces of Prussia as a result of emigration and the so-called "Ostflucht".[23] These movements generally affected the population of German origin more than that of Poles, who remained in the areas they had inhabited.

Extreme values can be found in 1879 and 1909 when accordingly, 461 and 338 children were born. However, considering the legal status of children, the highest number of state-recognized children was born in 1875 – 432, and the least in 1909 – 326. The highest number of illegitimate children was born in 1889 – 38, and the lowest in 1905 – only 5.

Also, in terms of percentage, the highest number of illegitimate children – 9.38%, was born in 1889. The lowest number of illegitimate children was in 1905 when they accounted for only 1.96%. More non-marital births occurred until the mid-1890s. There was a marked decline in the early 20th century, with a spike in 1910. On average, non-marital births account for 6.13%. Overall, however, there is a downward trend throughout the period under study.

Extramarital Births According to Religious Groups

The next problem is how the distribution of children from non-marital unions looked for each religious group. This question will be analyzed in the following section. Chart 8 shows the total fertility rate by religion.

The cumulative data show that Protestant denominations predominated among the rural population in the study area, which was reflected in the number of births, which, apart from 1898 and 1903, was generally higher among Protestants. The total disproportion between Protestant births (7,895) and Catholic births (6,529) is 1,366. Only 41 Jewish births were registered during the whole research period. Therefore, the number of births corresponds to the data on the demographic structure of the districts under study.

---

23   Trzeciakowski, Relations, 1990, p. 181.

Figure 8: Total fertility rates by religion

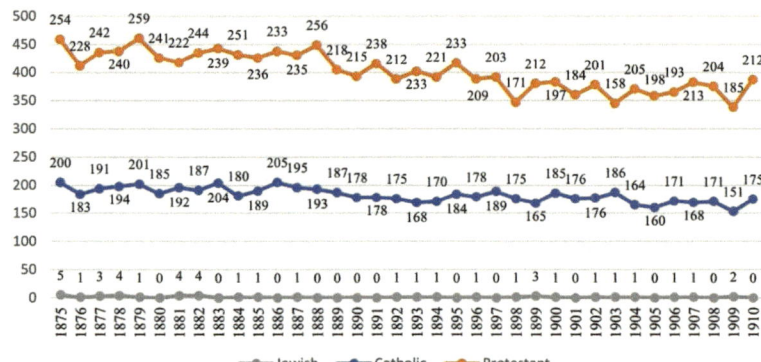

Source: own research

The trend of a decreasing birthrate can be observed among both Christian groups. However, the decline is more significant among Protestants than among Catholics. Among the Protestant population, a sharp drop in the number of births occurred from 1896 onwards. The decline in the number of births among Catholics, although pronounced, was significantly less.

The data on births is also worth analyzing in more detail regarding the various religious groups living in West Prussia.

Figure 9: Number of marital and extramarital births – Catholics

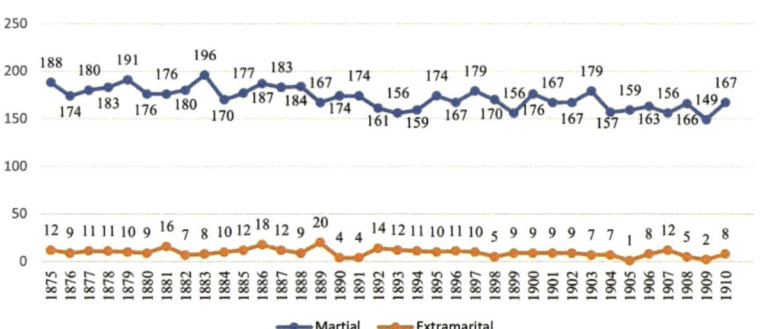

Source: own research

Figure 10: Percent of extramarital births – Catholics

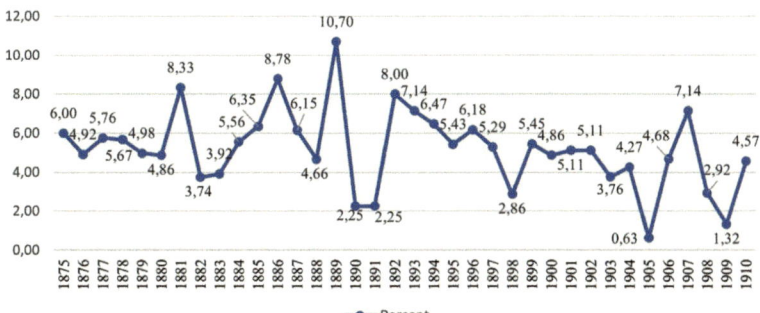

Source: own research

Charts 9 and 10 collect data on marital and non-marital births among Catholics living in the surveyed districts. The first chart presents data on absolute numbers. The second, on the other hand, shows percentage data. The number of illegitimate births among Catholics is relatively constant. In the entire period there were 341 extramarital births compared to 6,188 births from formal unions. The highest number of illegitimate births was 20 in 1889 and the lowest in 1905, only one. The percentage of children born out of wedlock was also the highest in 1889 (10.7%) and lowest in 1905 (0.63%). For the period as a whole, illegitimate births account for 5.22%. However, it is clear that until 1889 the percentage of extramarital births increased. Then, after two years of decline in 1890–1891, it returned to a relatively high level of 8% in 1892, before falling consistently for the rest of the period under study.

Figure 11: Number of marital and extramarital births – Protestants

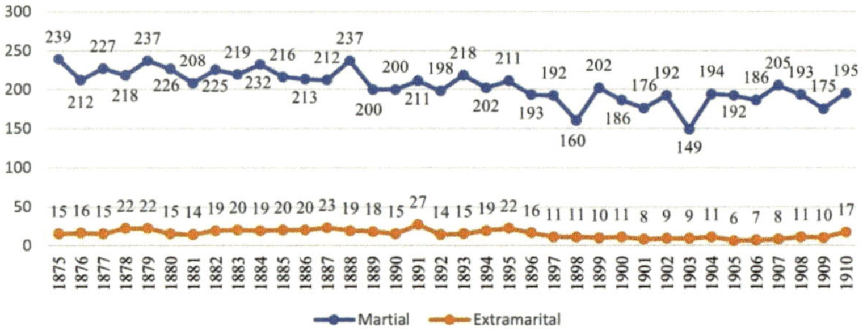

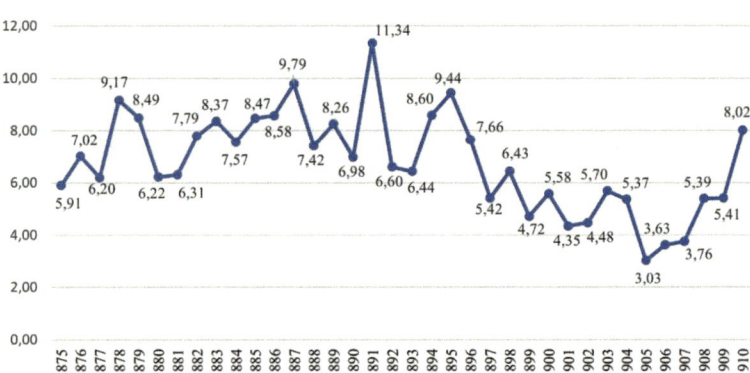

Source: own research

Figure 12: Percent of extramarital births – Protestants

Source: own research

The Protestant population shows a very similar pattern to the Catholic population. The number of legitimate-births is more or less constant with a slight downward trend. The number of extramarital births was stable up to 1896, except for 1891, when the number of extramarital births was higher than in the other years, amounting to 27. From 1897 onwards, there was a sharp decline, ending with an increase in 1910.

The percentage figures generally confirm the trend seen based on absolute numbers. They show, however, that the upward trend in the number of non-marital births begins again from 1905. This change appears to have been relatively stable, as an increase was recorded annually until 1910. Ultimately, 7,351 legal births and 544 extramarital births were recorded. Births from illegitimate marriages accounted for 6.89% of all recorded births.

Figure 13: Number of marital and extramarital births – Jews

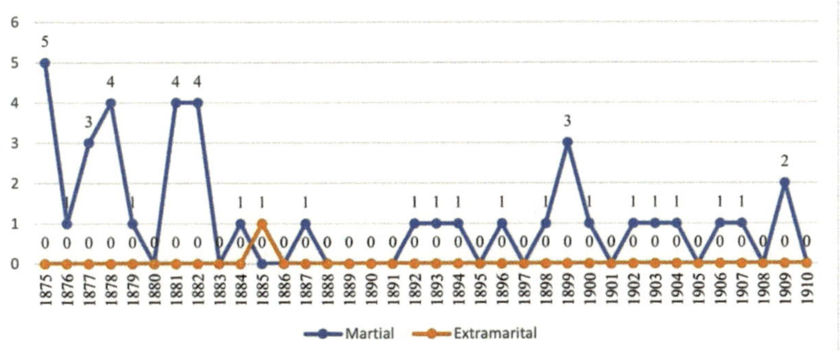

Source: own research

It is difficult to say anything definite about trends in the Jewish population due to the minimal amount of data collected. The highest number of births – 5, was registered in 1875. Interestingly, a child born out of wedlock was recorded only once in 1885. This single birth represents 2.44%. In this case, it makes no sense to analyze the percentage data.

Figure 14: Percent of extramarital births by religion

Source: own research

To compare the data presented, one must look at the percentages because of the differences in absolute numbers. In summary, presented on the chart above, the Jewish population was not taken into account,[24] because the number of illegitimate births was 0%, except for 1885, when it was 100%. However, returning to the summary, there is a clear predominance of illegitimate births among the Protestant population, with the exceptions of 1881, 1886, 1889, 1892, 1899, 1901–1902, 1906, and 1907. Apart from these years, the difference between the Christian denominations is relatively significant and amounts to 4 percentage points.

In summary, it should be stated that the most significant number of children from extramarital unions were born to mothers of the Protestant denomination (6.89%). Compared to Catholic women (5.22%), there were 1.67% percentage points more of them. In absolute numbers, it was 203 more children. The lowest number of illegitimate births was in Jewish families, only 2.44%.

---

24   For different activities of the Jewish population (but outside statistical data) see in this volume Izabela Spielvogel: The Jewish Women's League of Breslau, and Heidi Hein-Kircher: Debating Birth Control in Interwar Polish-Jewish Contexts.

Table 2: Marital and extramarital births according to the denomination

	Marital	Extramarital	Total	Percent
Catholic	6188	341	6529	5.22%
Protestant	7351	544	7895	6.89%
Jewish	40	1	41	2.44%
Total	13579	886	14465	6.13%

Source: own research

In West Prussia and Poland

To answer the question posed at the beginning of the paper, the historical data must be juxtaposed with another sample. To find some contrast, the author decided to compare the gathered data with the information about the fertility of the population of Poland in the period of the People's Republic of Poland and the Third Republic of Poland. This comparison is presented in Table 3.

Table 3: Per cent of extramarital births in Poland

Year	1875 - 1910	1950	1960	1970	1980	1990	2000	2010	2019
Percent	6.13%	6%	4.5%	5%	4.7%-4.8%	6.2%	12.1%	20.6%	25.4%

Source: P. Szukalski, *Płodność pozamałżeńska w Polsce*, Studia demograficzne 2(136), 1999, p. 111; Demographic Yearbook of Poland, Warsaw 2020 (https://stat.gov.pl/obszary-tematyczne/roczniki-statystyczne/roczniki-statystyczne/rocznik-demograficzny-2020,3,14.html, 05.06.2025), p. 250

As can be seen, until 1990 in formally secular, communist Poland, in which the state placed great emphasis on shaping attitudes and traditions detached from the Church's teachings, the percentage of children born out of wedlock was lower than in 1875–1910. It was not until the end of the 20th century and the

21st century that the number of out-of-wedlock births increased significantly, even by leaps and bounds.

Conclusion

What conclusion, then, can be drawn on this basis? The Protestant population appears to have more extramarital partners than the Catholic population based on the random sample surveyed. The difference between the Christian denominations is 1.67 percentage points. The lowest number of extramarital births was recorded among the Jewish population. However, it is challenging to analyze this data due to the decidedly small number of recorded Jewish births in general. Of course, the author is aware of the limited nature of the sample, but it seems to provide a good starting point for further research on this topic.

The second conclusion can be found by comparing 1875–1910 and the period after the Second World War. The picture emerging here shows that the population circa 1900 was no less susceptible to the temptations of the flesh than subsequent generations up to the turn of the next century. So the close bond with churches and religiosity was not so important when it came to intimacy between men and women.

The preliminary research presented in this article opens up the field for further research into the actual state of the influence of society's religiosity at the turn of the 20th century on the sexuality of the population, which will provide insight into the actual state of the churches influence on the morality of the population during the period under study. This research will also be a prelude to studying changes in this area in other historical periods. The results presented here are thus only the first stage and an exemplification of the possibilities offered by statistical research based on marital status registers kept by the state, and therefore devoid of a religious component. Of course, this is only one of the many possibilities opened up by the broader use of these historical sources.

Bibliography

Gemeindelexikon für die Provinz Westpreussen. auf Grund der Materialien der Volkszählung vom 1. Dezember 1905 und anderer amtlicher Quellen, Verlag des Königlichen Statistischen Landesamts, 1908.

Gemeindelexikon für die Provinz Westpreussen. auf Grund der Materialien der Volkszählung vom 1. Dezember 1885 und anderer amtlicher Quellen / bearb. vom Königlichen statistishen Bureau, Verlag des Königlichen statistischen Bureaus, 1887.

CHWALBA, Andrzej, Historia Polski 1795–1918, Kraków, Literackie, 2000.

CLARK, Christopher, Prusy. powstanie i upadek 1600–1947. Christopher Clark, przeł. z ang. Jan Szkudliński., Warszawa, Bellona, 2009.

FRACKOWIAK, Johannes: Eine longue durée polnischer Ethnizität? Polen in Mitteldeutschland von 1880 bis zur Gegenwart, [In:] Deutsche und polnische Migrationserfahrungen. Vergangenheit und Gegenwart, ed. Dorota Praszałowicz, Frankfurt am Main, Bern, Bruxelles, New York, Oxford, Warszawa, Wien 2014.

KUCHARCZYK, Grzegorz, Prusy, pięć wieków, Warszawa, Bellona, 2020.

MORISON, John (ed.), Eastern Europe and the West. Selected Papers from the Fourth World Congress for Soviet and East European Studies, Harrogate, 1990, 1st ed. 1992., London, Palgrave Macmillan UK, 1992.

POCIECHA, Józef, The Process of Demographic Transition in Lands of the Former Polish-Lithuanian Commonwealth and Other Areas with Polish-Speaking Populations, 1865–1912, in: Przeszłość Demograficzna Polski, 42 (2020), p. 123–146.

SZUKALSKI, Piotr, Płodność pozamałżeńska w Polsce, in: Studia Demograficzne, 136 (1999), p. 109–123.

TRZECIAKOWSKI, Lech, Relations between the Polish and German Population of Prussian Poland, 1772–1918, in: Eastern Europe and the West. Selected Papers from the Fourth World Congress for Soviet and East European Studies, Harrogate, 1990, ed. John MORISON, 1st ed. 1992, London, Palgrave Macmillan UK, 1992, p. 173–185.

The *Jewish Women's League* of Breslau
Its Efforts to Protect Reproductive Health and the Health of Women and their Children in the Early 20th century

Izabela Spielvogel[1]

Abstract *The Breslau (now Wrocław) branch of the Jewish Women's Association (Jüdischer Frauenbund Ortsgruppe Breslau) was founded on October 6, 1908. The association grew out of the tradition of Jewish charitable organizations and was based on programs that the pioneers of the movement proposed as a women's conception of how to approach their place in the Jewish community. The issue of the contribution of Jewish women in Silesia to the field of reproductive health at the beginning of the 20th century has not yet received a thorough academic synthesis or analytical study. The current state of research on the subject contains many areas of unrecognised or silence and is far from being considered satisfactory. This article is therefore an attempt to fill this gap and to contribute to the discourse undertaken in the presented post-conference volume. The paper focuses primarily on an attempt to reconstruct the activities of the Jewish Women's League in Silesia – the leading organisation for Jewish social and medical care in the early 20th century, operating in the region for three decades. In addition to reconstructing the League's contribution to reproductive, perinatal and women's and their children's health in the Silesian Jewish community, the aim of the article was to contribute to the multidisciplinary discourse in areas related to, among others, the history and sociology of medicine, politics or law, centred on the issue of reproductive health. The following questions were therefore posed as part of the research: What were the specific problems and challenges in the health care of Jewish women in Silesia at the beginning of the 20th century? What was the process of organising social assistance and medical care for Jewish women in the region? How did the activities of the Women's League affect the situation of Jewish women in Silesia at that time?*

---

1 Department of Physiotherapy, Opole University of Technology.

Keywords: *health care; reproductive health of Jewish women; Jewish Women's League; Breslau (Wrocław); Silesia/Śląsk; first half of the 20th century*

Introduction

The activity of Jewish women in Silesia (*Śląsk* in Polish) in the first half of the 20th century in the field of medical care in the and context of reproductive health was an important area of their pro-social activity. The changes occurring in nineteenth-century Prussian society and the rich and multifaceted Jewish tradition were important for this activity. Judaism integrated moral-religious and hygienic precepts into one coherent system, which indicated, among other things, the importance of taking care of one's own health, as well as caring for the sick, handicapped and crippled.[2] The turn of the 19th and 20th centuries brought the greatest activity of the Jewish community in the public space, also in Silesia. This process was particularly evident in Breslau (schlesisch *Brassel*, yid. *Bresle*).[3] Organizations established in the city were of varying character: from religious associations, to political, economic, and cultural ones. With the emancipation act of 1812, which brought equal rights for the Jews, the abolition of the last inequalities in the documents of 1847–1848, and the granting of full civil rights in the Reich Constitution of 1871, a true golden age began for the Jews in Germany. Through their activities in associations and foundations, which grew out of the Judaic *mitzvah* of *tzedakah*, or commandment to charity (the word *mitzvah* comes from the Hebrew *cawo*, meaning "command," and the word *tzedakah* comes from the Hebrew word for justice, but is most often translated as "charity"), women made a significant contribution to the development of social and medical care, including reproductive and perinatal health in the region. Additionally, and this is important to remember, the command to provide relief to the sick and needy in Judaism is not a voluntary matter, but an obligation flowing from the Pentateuch: "you shall care for the poor, the sick, the widow, the orphan".[4]

---

2 Spielvogel, Spałek, Procków, Jewish doctors 2018, p. 680–685.
3 Dylewski, Śladami Żydów Polskich, 2019, p. 404. Beider, Toponyms, 2012, p. 449.
4 See: Babylonian Talmud: Sanhedrin 19. Malzerowa, Istota i zagadnienia opieki społecznej, "Przegląd Społeczny Miesięcznik poświęcony zagadnieniom pracy społecznej i opieki nad dzieckiem. Organ Związku Towarzystwa Opieki nad Żydowskimi Sierotami we Lwowie" 1927, J. 1, no.1, p. 5.

Jewish social organizations operating in the field of basic health care in Germany – including Silesia – had a particularly long tradition: there were many of them that dated back to the twelfth century.[5] This experience made it possible to develop a model for the functioning of medical institutions such as hospitals, nursery schools, dispensaries, treatment and prevention centers in health resorts, and to gather a significant part of the society around the issue of pro-health education. Among the Jewish philanthropic associations in Germany before 1945, there were more than two thousand societies involved in health care as part of social welfare with a total of more than two hundred thousand members.[6] It should be noted that the German Jewish population at that time numbered about half a million people.[7] Jewish organizations took care not only of their co-religionists, but also of the other inhabitants of the localities in which they were active. Paradoxically, the scope of Jewish social protection in Germany far exceeded its needs. Thus, before World War I, charitable activities were directed toward Jewish immigrants from Eastern Europe, migrating through Germany to other countries – mainly England and the United States, but also toward non-Jews in need.[8]

In the context of these organizations, the *Jewish Women's League* – *Jüdischer Frauenbund* (J.F.B) – presented itself as one of the most serious and largest associations with four hundred and thirty branches and eighty thousand members in pre-war Germany.[9] During the first ten years, thirty-five thousand women joined the organization. By 1929, there were 430 branches and 34 chapters with a total of fifty thousand members, accounting for more than 25 percent of Jewish women in Germany over the age of thirty. The *Jüdischer Frauenbund* (J.F.B) was founded in Berlin[10] in 1904 at the initiative of courageous Jewish suffragettes: Bertha Pappenheim (1859–1936)[11], Sidone

---

5   Herzig, Jüdisches Leben, 2018. Tatarkower, U podstaw żydowskiej opieki społecznej, 1931, p. 385–397.
6   Tartakower, Żydowska ochrona społeczna w Niemczech hitlerowskich, "Przegląd Społeczny. Miesięcznik poświęcony zagadnieniom pracy społecznej i opieki nad dzieckiem. Organ Związku Towarzystwa Opieki nad Żydowskimi Sierotami we Lwowie" 1936, J. 10, no 4–5, p. 76.
7   Herzig, Jüdisches Leben, 2018.
8   Ibid. p. 78.
9   Ibid. p .79.
10 Cf. Dämmig, Klapheck, Debora's Disciples, 2006.
11 Born in Vienna, an emancipationist, social activist, and poet. In 1881 she moved to Frankfurt am Main, where she became involved in the German emancipation move-

Werner (1860–1932),[12] who came from Poznań (Poland), and Henriette May (1862–1928).[13] Its foundations rested with Judaism and Jewish moral precepts, but the organization's goal was primarily to represent the interests of Jewish women in Germany, to improve their position in society, to activate women, and to strive for equality understood as a social alliance, rather than revolutionizing society. The organization's goals also included: education in various fields, professional preparation (especially in the areas of medical care and education), health care, combating prostitution, and the trafficking of Jewish girls. The statutory goals of the organization focused, among other things, on promoting and supporting the idea of women's equality in public life, access to suffrage, but also assistance in confinement, and basic assistance in reproductive and perinatal health. These demands were particularly reflected in social welfare and medical projects, which emphasized the important role of women in society and the idea that simple and practical reforms would improve the position and living conditions of women. As a result, care and treatment places were created for unmarried mothers condemned by society, as were centers for Jewish girls at risk of prostitution, kindergartens for children of single women who wanted to work, as well as help from midwives and community nurses in caring for newborns. The Association published a cookbook taking into account the dietary principles of Judaism and containing recipes and dishes to be served to children and sick adults.[14] The organization senior to the Jewish Women's Association was the All-German Association of German Women's Associations (*Bund Deutscher Frauenvereine*) founded in 1894

---

ment. Between 1904 and 1924 she was the president of the Jewish Women's League. She died in 1936 in Neu-Isenburg. Cf. Kozińska-Witt, Ostjüdinnen, 2011, p. 69–87.

12   A native of Poznań, she was an emancipist, educator and social activist. In 1893, together with Gustav Tuch (1834–1909), she founded the Israeli Humanitarian Women's Association (Israelitisch-Humanitärer Frauenverein), which she chaired from 1908 to 1932. She died in Bad Segeberg in 1932. Cf. Weissberg, Frauenbund, 2016, p. 138.

13   A native of Berlin, educator and social activist, graduate of a teachers' seminary. In the Jewish Women's Association, she served as secretary and member of the board. The first woman to serve on the board of the Central Verein Deutscher Staatsbürger Jüdischen Glaubens (Central Association of German Citizens of the Jewish Faith), which she co-founded (1918). She died in Berlin on May 14, 1928. Cf. Weissberg, See Footnote 166, p. 143.

14   Kochbuch, 1926, p. 230–239, 242–243.

on the initiative of Auguste Schmidt (1833–1902)[15] from Breslau, who headed the organization until 1899.

The *Jewish Women's League* of Breslau and its Activities in the Field of Reproductive Health and the Health of Women and their Children

The Breslau branch of the *Jewish Women's League* (*Jüdischer Frauenbund Ortsgruppe Breslau*) was founded on October 6, 1908. This date is significant because on October 1 of that year women in Prussia were permitted to study at universities for the first time. The structure of the organization was based on branches that concentrated on work in different areas. The activities of the Silesian Jewish women breathed freshness and practicality under the new challenges of industrial society for women.[16] The organization's goals, which focused broadly on issues related to protecting the health of mothers and their children, including reproductive and perinatal health issues, were pursued primarily through the Health Care Branch (*Erholungsfürsorge*) and the Girls' Branch (*Mädchenklub*). During World War I, many displaced, poor Jewish girls from the East, mainly from Galicia (formerly part of the *Austro-Hungarian Empire*, now part of Ukraine), were taken care of. It became an important task for the branch to take care of them in order to prevent them from falling into the clutches of prostitution. The direction of the Association's activities in the context of taking care of girls concerned a courageous and open discussion about Jewish prostitution.[17] Apart from poverty, the founders of the association pointed to the legal status of women in Judaism as one of the causes of Jewish prostitution, claiming that Jewish divorce laws significantly contributed to the increase in the number of women engaging in prostitution. During World War I, many men left their homes and never returned. Women remained *agunot*, or "wives chained to their marriage," unable to remarry without an official

---

15 A native of Breslau, she was a pioneer of the women's emancipation movement in Germany. In 1865 together with Louise Otto-Peters (1819–1859) she founded the Allgemeiner Deutscher Frauenverein (ADF) in Leipzig. In 1890, together with Helene Lange (1848–1930), an educator and suffragist, she founded the Allgemeinen Deutschen Lehrerinnen-Vereins (ADLV). In 1894 she became the first president of the League of German Women's Associations (Bund Deutscher Frauenverein – BDF), which she founded. She died in 1902 in Leipzig. Cf. Plothow, Begründerinnen, 1907.
16 Ibid. p. 449.
17 Pappenheim, Rabinowicz, Verhältnisse, 1904, p. 76.

Jewish divorce, which requires the man to divorce his wife, not the other way around, or demands Jewish witnesses to the husband's death.[18] The prevention of venereal disease was also an important part of the Association's activities. Therefore, as part of the work of the branch focused on health, comprehensive care, education, and assistance were undertaken for girls as well as single women and pregnant minors. The branch also developed activities around adoption mediation, which yielded extremely good results. The phenomenon of mass orphanhood, which emerged in the years 1914–1918, forced the Jews to intensify their activities within the modern system of care. In the post war years, it became the basis for further development of orphan action. After 1916, the Breslauer Home for Infants and Children of the *Jewish Women's League* (*Breslauer Säuglings -und Kleinkinderheim* des J.F.B.), which comprehensively supported women extended its services to take care of their young children.

Preventive and Perinatal Actions

One of the important goals of the work in the area of perinatal health was preventive measures that focused on education and the promotion of healthy lifestyles among members (which took place under the guidance of a doctor). The work specifically focused on strengthening the role of young women in the Jewish community. This goal was pursued by providing girls with access to health-promoting education and vocational training that would give them independence. For professional work was seen as a condition for economic, psychological, and emotional independence. Therefore, the *Jewish Women's League* supported and organized professional training for poor girls threatened with prostitution, established employment offices and vocational counselling centers. The first meeting of the Breslau girls' section of the *Jewish Women's League* took place on October 26, 1913 in modest rooms at the then Neue Schweidnizer Strasse 10 (now Świdnicka Street).[19] Lisbeth Cassirer née Lasker (1886–1974),[20] known for her love of art, became the chairperson of

---

18   H. Kozińska-Witt, Ostjüdinnen, 2011, p. 69–87.
19   Specifically, its section from the Moat and the intersection with Podwale to the railroad viaduct (author's note).
20   She was a co-editor of the Calendar of the Jewish Women's League published every year. In the 1930s she moved with her husband from Breslau to Berlin, where in 1934 she founded the local branch of the Jüdischen Kulturbund (Jewish Art Association). During the Nazi era, her Berlin apartment functioned as an art salon, due to the ban-

the section. The supervisors of the Breslau group in particular years were: Ina Heimansohn 1913–1922, Dora Hirschberg 1923–1926, Erna Stein-Blumenthal 1925–1926, Margarete Danziger 1926–1927, Qara Müller 1927–1928.[21] The founding group consisted of thirty people. That same year, the section began organizing continuing education classes on topics such as health care. In 1916 the seat of the section was moved to the bigger and more comfortable rooms in Agnesstrasse (now Michała Bałuckiego Street). In May 1918 the section, which at that time numbered 114 members, was moved to new, still bigger rooms at Freiburger Strasse 15 (now Świebodzicka Street 15). From 1918 the meetings of the Breslau youth section were held once a month at first, and then more often (even up to 4–5 times a week).

During the war many displaced, poor Jewish girls from the East joined the youth unit. It became an important task of the unit to take care of them so that they would not go down the road of prostitution. Decisions concerning this issue were made in October 1928 in Breslau, during the general convention of the *Jewish Women's League*. The issue provoked lively discussion and concern, along with a declaration of international cooperation. The leading activists of the Polish branch of the *Jewish Women's League* – Róża Malzerowa and Ada Reichenstein[22] of Lviv – wrote about the problem in 1928 e.g., in the pages of the Social Review: "the percentage of Jewish prostitution from Poland is relatively

---

ning of Entartete Kunstv – degenerate art – for many Jewish artists. In 1938 she left Nazi Germany and went to London with her husband and daughter Susan Cf. Bauschinger, Cassirers, 2015. Stein-Blumenthal, Geburtstag, 1966, p. 11.

21 E. Rabin (ed.), Gedenkbuch Jüdischer Frauenbund Ortsgruppe Breslau, Breslau 1928, p. 62.

22 Ada Kalmus Reichenstein (1880–?) was born in Stanislawow. She defended her doctoral dissertation in philosophy (on the basis of Das moderne Märchendrama) in 1903, as one of the first women students at Lviv University. Her husband was a doctor Marek Reichenstein, (1867–1932) an assistant in the Clinic of Internal Medicine at the Medical Faculty of Lviv University, known for his love of art history, co-founder of the Jewish Museum in Lviv and owner of the largest collection of Judaica in that city before the war. Ada Kalmus Reichenstein was also the head of the committee for building a sanatorium "Jewish Academy" in Vorokhta. Cf.: Ł.T. Sroka, Stowarzyszenie Humanitarne "Leopolis" we Lwowie (1899–1938), "Kwartalnik Historyczny" 2016, no. 123, p. 63. Ada's sister was Maria Kalmus Schneiderowa – the first woman in the history of Lviv University to study medicine – she became a gynecologist. Both sisters were famous for their pro-social and charitable activities. They belonged, among others, to the board of the Jewish Women's League in Lviv. See: From the Editorial Board, Z Towarzystwa Ochrony Kobiet, "Kurjer Lwowski" 1911, J. 29, no. 151, p. 6.

enormous".²³ We face a whole range of issues such as: "[...] the care of the illegitimate child, the abandoned and homeless woman, the fight against fornication, the fight against human traffickers [...]".²⁴ Meeting the basic needs of immigrant women, such as help with organizing life in the city – but also psychological care, medical care, and education in the context of reproductive health – became a priority. To this end, an apartment was rented in Breslau in Brüderstrasse (now Bracka Street), where the girls were accommodated. As part of the section's activities, further education and vocational courses were introduced, e.g., in the field of medical care. Initially, the education programs took place mainly in the evening mode and concerned such issues as: basic health care, venereal diseases, as well as pediatric nursing and social work with sick children, e.g., blind or deaf-mute.²⁵ After completing courses, the girls were sent to hospitals or children's centers where they helped with nursing or organized various forms of therapeutic activities (e.g., books were read to the blind and sick). In addition, the Association helped to place girls with illegitimate children in the Home for Single Mothers of the Jewish Women's Association in Isenburg by transferring monthly funds.²⁶ At that time, public opinion – both Jewish and Christian – was particularly strong in its condemnation of unmarried mothers and their illegitimate children. The center protected women who cared for their offspring.

Another important area of support for young women in need was to help them stabilize their personal lives by providing them with the necessary means to get married. The obligation (*mitzvah*) to support poor Jewish women

---

23   R. Malzerowa, W odpowiedzi p. Dr Reichensteinowej, "Przegląd Społeczny. Miesięcznik poświęcony zagadnieniom pracy społecznej i opieki nad dzieckiem. Organ Związku Towarzystwa Opieki nad Żydowskimi Sierotami we Lwowie" 1928, J. 2, no. 6, p. 24.
24   A. Reichenstein, O międzynarodowej, międzypartyjnej i międzywyznaniowej pracy społecznej, "Przegląd Społeczny. Miesięcznik poświęcony zagadnieniom pracy społecznej i opieki nad dzieckiem. Organ Związku Towarzystwa Opieki nad Żydowskimi Sierotami we Lwowie" 1928, J. 2, no. 11, p. 3.
25   Editor's note, Und welche Lehre für das Mädchen, "Jüdisches Gemeindeblatt" 1938, Jg 15, Nr. 4, p. 2.
26   The center was founded in 1907 in the state of Rhineland-Palatinate. It was the first home for single mothers in Germany. Cf.: B. Pappenheim, Aus der Arbeit des Heimes des J.F.B. in Neu-Isenburg 1924–1929, "Blätter des Jüdischen Frauenbund. Das Sonderdruck" 1930, Jg 6, no. 1, p. 3; L. Daemmig, M. Kaplan, Jüdischer Frauenbund (The League of Jewish Women), Jewish Women: A Comprehensive Historical Encyclopedia. 27 February 2009, Jewish Women's Archive. https://jwa.org/encyclopedia/article/juedischer-frauenbund-league-of-jewish-women (05.06.2025).

wishing to marry – *Hachnossas Kalloh*[27] – has always ranked among the most sacred duties which no one could evade. During the period of the organization's existence, which spanned two world wars, this injunction gained special importance in the community. The problem for Jewish women was the "spreading threat"[28] in the form of disgraced child brides and mixed marriages that called into question the continued fate of Judaism. Faced with this fact, in February 1927 the *Jewish Women's League* set up a relief fund for future married women. Help was manifested both in the form of regularly paid membership fees and donations of household equipment and furnishings. Support was also provided in the form of assistance in organizing wedding ceremonies or dowries for indigent girls, which were requested from well-to-do members of the Association. By 1928, the fund had made it possible for eighteen young ladies to get married and build a house.[29]

House for Young Children and Infants in Breslau

The House for Young Children and Infants (*Breslauer Kleinkinder und Säuglingsheim des J.F.B. in Krietern*) was founded in 1916 in the Krietern (now Krzyki) district as a home for Jewish orphans and children of single mothers who lacked sustenance. Judaism has always regarded the care of orphans as a task of great importance. The question of orphans in Breslau became extremely urgent to solve from the beginning of World War I. Under the conditions of the war, an important issue arose to create a place for the poorest, neglected, abandoned, and most often illegitimate children.[30] Because of their "illegal" statute, these children were excluded from orphanages and educational homes intended for school children. For this purpose, the Jewish community in Breslau granted a subsidy. With a contribution of 4,276 marks, a house with a garden in Moritzstrasse (now Lubuska Street) was rented. The center was also maintained thanks to the support of the American Joint Distribution

---

27 Cf.: A. Simonsohn, Ausstattung für Bräute, [in:] E. Rabin (ed.), Gedenkbuch Jüdischer Frauenbund Ortsgruppe Breslau, Breslau 1928, p. 42; P. S. De Vries, Obrzędy i symbole Żydów (Jewish rites and symbols), A. Borkowski (transl.), Kraków 2001, p. 292.
28 A. Simonsohn, op. cit., p. 42.
29 Ibidem, p. 43.
30 J. Cohn, Breslauer Kleinkinder und Saeuglingheim des J. F. B. in Krietern, [in:] E. Rabin (ed.), Gedenkbuch Jüdischer Frauenbund Ortsgruppe Breslau, Breslau 1928, p. 20.

Committee.³¹ The children's health was supervised by a pediatrician, Dr. Franz Steinitz (1876–1931),³² who monitored their health and aimed at preventive measures. In 1928, Else Toeplitz, the caretaker and manager of the center, took over the supervision of the children. At that time, the ward chairwoman, Paula Ollendorff,³³ and Johanna Cohn, the wife of the sanitary councillor Richard Cohn,³⁴ were responsible for all the center's affairs. In the first quarter of 1918, the center housed 32 children. By 1928, 165 little wards found accommodation for longer and shorter stays; 16 of them were adopted. In 1919, a part of the invested capital was used to buy a beautiful secluded garden in the Krietern district of Breslau. In the summer of 1925, due to the unsatisfactory condition of the building, the center changed its address to Trentinstraße 35 (now Krzycka Street 35), where it occupied a building that met the latest hygienic standards of the time. In July 1927, thanks to numerous donations, the center established a ward for infants.³⁵ Thus, professional health care was undertaken for infants and children who were considered particularly vulnerable

---

31   The organization was founded on November 27, 1914 on the initiative of the American Jewish Committee. It was supposed to help Jews in war-stricken Europe by distributing funds collected by charitable organizations: American Jewish Relief Committee, Central Relief Committee and People's Relief Committee. The organization still exists today, providing assistance to Jewish communities in need. Cf.: J. Tomaszewski, A. Żbikowski (ed.), Żydzi w Polsce. Dzieje i kultura. Leksykon, Warszawa 2001.

32   Editor's note, Die Hauptversamlung der Breslauer Ortsgruppe des J. F. B, "Breslauer Jüdisches Gemeindeblatt" 1931, Jg 8, nr 12, p. 168.

33   She was born on May 18, 1860 in Kostomłoty near Breslau (then Kostenblut). In Breslau she graduated from a teachers' seminary and worked as a teacher in Budapest and London. She was a prominent social democratic politician, and in 1918 she was the first woman to be elected to the Breslau city council. While holding this position, she founded many Jewish charitable organizations, engaging in charitable care herself. She was a co-founder of the Jewish Welfare Office in the Breslau municipality. In 1920 she became the chief chairman of the Jewish Women's League and was also active on the board of the World Union of Liberal Judaism. She always felt part of the German people and considered the Zionist movement the wrong way to go. She spoke several languages, was a well-read person, and took an active part in the cultural life of Breslau. In 1937, at the invitation of her son Friedrich Ollendorff (who was a Zionist), she went to Palestine. She fell ill there and died in 1938. She was buried in the cemetery on the Mount of Olives. Cf.: H. Vogelstein, Paula Ollendorf zum Gedächtnis "Jüdisches Gemeindeblatt für die Synagogen-Gemeinde Breslau" 1938, no. 20, p.1.

34   E. Rabin, op. cit., p. 61.

35   E. Landsmann, Das Kinderheim des Jüdischen Frauenbundes Breslau, "Breslauer Jüdisches Gemeindeblatt" 1927, Jg 4, Nr. 8, p. 119.

to disease. The organization thus played a pioneering role in the development of pediatric medical care (modern forms of intervention and innovative measures). Care was taken to provide isolation rooms for sick children with separate toilets for them, washable walls and floors, separate entrances from the hallway and staircase, and access to running hot water. In order to reduce infant mortality, mechanisms of infant health control and prevention were developed in accordance with the state of knowledge at that time.

Nursing Assistance

Care for the sick in Judaism derives from the commandment to visit the sick (hebr. *Bikur Cholim*, yid. *biker-chojlim*), which carries the broader sense of nursing them. According to the Talmud (Nedarim 39b), visiting an ill person renders him or her sixty percent sick, and failure to visit may lead to the death of one's neighbour. Fulfilling this order in Jewish communities was the responsibility of special confraternities (hebr. *chewot*) such as *Bikur Cholim*. Their members visited the sick and took care of their needs. Nursing in Judaism is thus connected with the command to visit the sick, constituting one of the most important precepts of Judaism and Jewish social ethics, since it serves to preserve and sanctify life. It is the "sacred duty" of both Jewish men and women.[36] In the 19th century, the nature of most training institutions (even with explicit Christian ideals) prevented Jewish women from participating in the nursing profession, so Jewish hospitals in Germany began to organize the first nurse training courses themselves. The first professional Jewish nurse in Germany is believed to be Rosalie Jüttner, who most probably came from Poznań (Poland) and was employed in 1881 in a hospital in Hamburg. In Breslau, nursing courses began in 1884 at the Fränckel Jewish Hospital on what is now St. Anthony Street. In 1899, the German Association of Jewish Nurses (*Deutscher Verband jüdischer Krankenpflegerinnen*) was established in Breslau. As for nursing help within the structures of the Jewish Women's Association in Breslau, they worked in all the institutions of the organization, and there were also community nurses who joined the group on behalf of the community in April 1914. Their tasks in the field of reproductive health included assistance in childbirth, coordination of puerperal home care, assistance in pediatric care, assistance in newborn

---

36   Cf.: H. Steppe Den Kranken zum Troste und dem Judenturm zur Ehre. Zum geschichte der judischen Krankenpflege in Deutschland, Frankfurt a. Main 1997.

care, care in case of puerperal infection, organization of visits to the doctor or purchase of medicines. In the first year of operation, 159 people received ambulatory care for various cases, including those related to childbirth and puerperium, and this number increased from year to year. In 1923, the number of members of the Breslau Jewish community covered by home nursing care amounted to 1,495 people, and in 1927 there were 2,492 outpatient cases. In 1928 the community nurse of the Association was Rosa Schönfelder.[37]

Therapeutic and Recreational Support for Women and their Children

The center belonging to the Breslau branch of the *Jewish Women's League* was a sanatorium for Jewish children in Bad Fliensberg (now Świeradów-Zdrój) (*Kinderlandheim des J.F.B. Bad Flinsberg*), which started operating in 1924 in the boarding house Loreley (currently at 6 Bronisława Czecha Street).[38] The center was financed, among others, from a grant by the Joint Distribution Committee and from a long-term loan taken from the *Hilfskasse gemeinnütziger Wohlfahrtseinrichtungen in Deutschland*.[39] Stays in health resorts were a form of assistance for the poorest or single-parent families, where the upbringing of children rested on single mothers. Before the establishment of the center, the conditions in which the children were placed were not always comfortable, e.g., during World War I children were accommodated in unused rooms of inns.[40] Emma Vogelstein née Kosack (1870–1949), wife of the liberal rabbi from Breslau – Hermann Vogelstein (1870–1942) – was the director of the Association of Women and responsible for all organizational matters from the beginning of the sanatorium's existence. She was born in Warglitten – East Prussia (now Warglity) and died in New York.[41] Children staying at the center were consulted by doctors (e.g., from 1924 to 1928 by Dr. Arthur Schäfer from

---

37 L. Cassirer, Jüdische Tuberkulosenfürsorge eine arbeitsgemeinschaft des Jüdischen Wohlfahrtsamts und des J.F.B., [in:] E. Rabin (ed.), Gedenkbuch Jüdischer Frauenbund Ortsgruppe Breslau, Breslau 1928. p. 53.
38 Cf.: Editor's note, Die Arbeit des Jüdischen Frauenbundes in Breslau, "Breslauer Jüdisches Gemeindeblatt" 1925, Jg 2, nr 11, p. 168.
39 E. Vogelstein, Rabin (ed.), op. cit., p. 30.
40 Editor's note, Besichtigung des Kinderlandheimes in Flinsberg, "Breslauer Jüdisches Gemeindeblatt" 1929, Jg 6, nr 9, p. 154.
41 W. Röder, H.A. Strauss (ed.), Biographisches Handbuch der deutschsprachigen Emigration nach 1933, Band I, Politik, Wirtschaft, Öffentliches Leben, Leitung und Bear-

Świeradów-Zdrój), and in the years 1937–1938 by Dr. Lucie Baas, a doctor at the Jewish hospital in Breslau. The house was run by a nurse with appropriate qualifications and who passed a state exam (e.g., in 1928 this was Selma Schloss). Small patients were looked after by educated tutors of both sexes (boys – men, girls – women) and assistants. Each tutor was responsible for a group of 10 to 12 children. The kitchen was run strictly according to ritual regulations, which was taken care of by the head of the kitchen, Josefine Kantorowicz. Apprentices helped with the cooking, thus supplementing their practical training. Consultants and committee members for the sanatorium included: Dr. Erich Breslauer, Lisbet Cassirer, Dr. Alfred Cohn, Richard Ehrlich, Gustav Glaser, Beate Guttmann, Guido Neustadt, Eugen Ollendorff, Eugen Perle, Siegfried Preuss, Ema Rosenberg, Leo Smoschower, Max Selberg, Anna Simonsohn, Clara Schottländer, Emmy Vogelstein, and Jacob Wolfssohn. The economic committee consisted of Helene Eichelbaum, Jenny Kochmann, Frieda Loebell, Friederike Löwenson, and Berta Schlesinger.[42]

The children stayed for an average of 28 days.[43] The average weight gain after the course was about 9 pounds (4.5 kg). An average of three hundred children were on holiday at the center during the year. Boys over the age of twelve were offered therapeutic stays in October.[44] From mid-December 1925, winter cure stays for children were also organized. The cost of these stays was three marks per day, two marks for the poorest children; however, it was stated that if someone paid four marks instead of three marks, he would be co-financing the stay for those who could not afford it. After the treatment in the health resort, the children were directed to the House for Young Children in Krietern, where therapy was continued in order to consolidate the therapeutic effects. An initiative of stationary recreational stays in the villa district of Kleinburg (now Borek)[45] was founded in 1928. During the summer season, 32 people usually stayed in a rented house with a garden. Most of them were women who had no chance to leave their place of residence for a longer period of time. Some of

---

    beitung, unter Mitwirkung von D. M. Schneider, L. Forsyth, S. C. B. Schmidt, München-London-Paris 1980, p. 784.

42 E. Vogelstein, Rabin (ed.), op. cit., p. 32.

43 Editor's note, Bezuch im Kinderheim Flinsberg, Die Arbeit des jüdischen Frauenbundes in Breslau, "Breslauer Jüdisches Gemeindeblatt" 1927, Jg 4, nr 10, p. 156.

44 Ibidem, p.157.

45 Currently a district of Wrocław, a village annexed to the city in 1897 thanks to the foundation of Julius Schottländer, who in 1877 bought the area of the Kleinburg villa colony and donated it to the magistrate. Cf.: A. Dylewski, op. cit., p. 413.

them, for example, could not stop working or were not able to raise enough money for a spa treatment. In addition to free rides, these women received afternoon tea and dinner.

Conclusion

In attempting to outline the final conclusions of own research, it should be emphasised that the pre-war activities of the Jewish Women League in Silesia in the areas of reproductive health and maternal and child health were part of the centuries-old traditions of activity of Jewish aid organisations. The League not only provided social and medical assistance, but above all became an important social movement for its time and for the women living in those times. This was made possible by a specific point of view derived from the precepts of Jewish social ethics, characterised by humanitarianism. The ideas of women's health care focused on the human being as a whole, and medical care was a comprehensive activity, far beyond medical procedures, and encompassing many areas of life. The bold projects of the Jewish Women's League highlighted the neglect and vital needs of women functioning in the society of the time, in which they were in a double minority – as women and as Jews. The simple and practical reforms implemented by the League under adverse conditions provided a biological sense of security and a basis of existence for several hundreds of women and their children. Their work resulted in the establishment of care facilities for unmarried or agunot mothers condemned by society, centres for Jewish girls at risk of prostitution, kindergartens for the children of these women who had to take up professional work to become economically independent. Educational assistance was also organised, e.g. towards becoming a midwife and nurse, in order to improve the care of newborns and mothers-to-be, also from the poorest families.

The activities of the Jewish Women League are also part of the complex landscape of Silesia, forever inscribed in the region's history, and perpetuating this heritage and opening up discussion on it contributes to defining regional and European identity.

Bibliography

APTROOT, Marion/GAL-ED, Efrat/GRUSCHKA, Roland/NEUBERG, Simon (eds.), Leket. Jiddistik heute, Düsseldorf, Düsseldorf University Press GmbH, (Jiddistik. Edition & Forschung, Bd. 1), 2012.

BAUSCHINGER, Sigrid, Die Cassirers. Unternehmer, Kunsthändler, Philosophen, München, C.H.Beck, 2015.

BEIDER, A., Eastern Yiddish Toponyms of German Origin, in: Leket. Jiddistik heute, ed. Marion APTROOT, Efrat GAL-ED, Roland GRUSCHKA und Simon NEUBERG, Düsseldorf, Düsseldorf University Press GmbH, (Jiddistik. Edition & Forschung, Bd. 1), 2012, p. 437–466.

BÜRO DER SYNAGOGENGEMEINDE (ed.), Und welche Lehre für das Mädchen? in: GSL P 28190 III, Jg. 15 (1938), Jüdisches Gemeindeblatt für die Synagogen-Gemeinde Breslau, p. 1–2.

CASSIRER, Lisbet, Jüdische Tuberkulosenfürsorge eine arbeitsgemeinschaft des Jüdischen Wohlfahrtsamts und des J.F.B., in: Gedenkbuch. Jüdischer Frauenbund, Ortsgruppe Breslau, ed. Else RABIN, p. 53.

COHN, Johanna, Breslauer Kleinkinder und Säuglingheim des J. F. B. in Krietern, in: Gedenkbuch. Jüdischer Frauenbund, Ortsgruppe Breslau, ed. Else RABIN, Breslau, 1928, p. 20.

DAEMMIG, Lara & KAPLAN, Marion, Juedischer Frauenbund (The League of Jewish Women), unter. Jewish Women's Archive, https://jwa.org/encyclopedia/article/juedischer-frauenbund-league-of-jewish-women, (05.06.2025).

DAEMMIG, Lara & KLAPHECK, Elisa, Debora's Disciples, A Woman Movement as an Expression of Jewish Renewal in Europe, in: Turning the Kaleidoscope. Perspectives on European Jewry, Oxford, Berghahn Books, 2006.

DYLEWSKI, Adam, Śladami Żydów polskich. przewodnik, reportaż, Bielsko-Biała, Pascal, 2019.

JÜDISCHER FRAUENBUND (ed.), Kochbuch für die jüdische Küche, Düsseldorf, herausgegeben vom Jüdischen Frauenbund,1926., 1926.

KOZIŃSKA-WITT, Hanna, Bertha Pappenheim und die Ostjüdinnen, Vol. 9 (2011), Scripta Judaica Cracoviensia, p. 69–87.

LUSTIG, Sandra/LEVESON, Ian (eds.), Turning the Kaleidoscope. Perspectives on European Jewry, Berghahn Books, 2006.

MELZEROWA, Róża, W odpowiedzi p. Dr. Reickensteinowej, in: Biblioteka Jagiellońska, 1073 II czasop., Jg. 2 (1928), Przegląd Społeczny. miesięcznik poświęcony zagadnieniom pracy społecznej i opieki nad dzieckiem, p. 20–26.

MELZEROWA, Róża, Istota i zagadnienia opieki społecznej, in: Biblioteka Jagiellońska, 1073 II czasop., (1927), Przegląd Społeczny. miesięcznik poświęcony zagadnieniom pracy społecznej i opieki nad dzieckiem, p. 5–8.

PAPPENHEIM, Bertha, Aus der Arbeit des jüdischen Frauenbundes, Jg. 6 (1930), Blätter des jüdischen Frauenbundes für Frauenarbeit und Frauenbewegung, offizielles Organ des jüdischen Frauenbundes von Deutschland, p. 3–10.

PAPPENHEIM, Bertha, RABINOWITSCH, Sara, Zur Lage der jüdischen Bevölkerung in Galizien, 1. Auflage., Frankfurt (Main), Neuer Frankfurter Verlag GmbH, 1904.

PLOTHOW, Anna, Die Begründerinnen der deutschen Frauenbewegung. Mit 24 Illustrationen, Leipzig, Rothbarth, 1907.

RABIN, Else (ed.), Gedenkbuch. Jüdischer Frauenbund, Ortsgruppe Breslau, Breslau, 1928.

RECHNITZ, Dr., Die Hauptversamlung der Breslauer Ortsgruppe des Jüd. Frauenbundes, in: GSL P 28190 III, Jg. 8 (1931), Breslauer Jüdisches Gemeindeblatt. Amtliches Blatt der Synagogengemeinde zu Breslau, p. 168.

RECHNITZ, Dr., Besichtigung des Kinderlandheimes in Flinsberg, in: GSL P 28190 III, Jg. 6 (1929), Breslauer Jüdisches Gemeindeblatt. Amtliches Blatt der Synagogengemeinde zu Breslau, p. 154.

RECHNITZ, Dr., Besuch im Kinderlandheim Flinsberg, in: GSL P 28190 III, Jg. 4 (1927), Breslauer Jüdisches Gemeindeblatt. Amtliches Blatt der Synagogengemeinde zu Breslau, p. 156–157.

RECHNITZ, Dr., Das Kinderheim des Jüdischen Frauenbundes Breslau. Breslauer Jüdisches Gemeindeblatt. Amtliches Blatt der Synagogengemeinde zu Breslau, 22. August 1927 Jg. 4 Nr 8, in: GSL P 28190 III, Jg. 4 (1927), Breslauer Jüdisches Gemeindeblatt. Amtliches Blatt der Synagogengemeinde zu Breslau, p. 119.

RECHNITZ, Dr., Die Arbeit des Jüdischen Frauenbundes in Breslau, in: GSL P 28190 III, Jg. 2 (1925), Breslauer Jüdisches Gemeindeblatt. Amtliches Blatt der Synagogengemeinde zu Breslau, p. 168.

REICHENSTEIN, Ada, O międzynarodowej, międzypartyjnej i międzywyznaniowej pracy społecznej, in: Biblioteka Jagiellońska, 1073 II czasop., (1928), Przegląd Społeczny. miesięcznik poświęcony zagadnieniom pracy społecznej i opieki nad dzieckiem, p. 3–6.

RÖDER, Werner/STRAUSS, Herbert A. (eds.), Biographisches Handbuch der deutschsprachigen Emigration nach 1933. Bd. 1. Politik, Wirtschaft, öffentliches Leben / Leitung u. Bearb. Werner Röder, Herbert A. Strauss unter

Mitw. von Dieter Marc Schneider, Louise Forsyth. Autoren. Jan Foitzik, München, Saur, 1980.

SIMONSOHN, Anna, Ausstattung für Bräute, in: Gedenkbuch. Jüdischer Frauenbund, Ortsgruppe Breslau, ed. Else RABIN, Breslau, 1928, p. 42–33.

SPIELVOGEL, Izabela, SPAŁEK, Krzysztof & PROĆKÓW, Jarosław, The Jewish doctors involved in the development of health resorts in eastern Galicia at the late 19th and early 20th century (Central and Eastern Europe), in: Wiener klinische Wochenschrift, 130 (2018), p. 680–685.

SROKA, Łukasz Tomasz, Stowarzyszenie Humanitarne "Leopolis" we Lwowie (1899–1938). główne kierunki działalności, in: Kwartalnik Historyczny R. 123 nr 1 (2016), 123, p. 45–69.

STEIN-BLUMENTHAL, Erna, Lisbeth Cassirer zum 80. Geburtstag, (1966), AJR Information, p. 11.

TARTAKOWER, Arjeh, Żydowska ochrona społeczna w Niemczech hitlerowskich, in: Biblioteka Jagiellońska, 1073 II czasop., (1936), Przegląd Społeczny. Miesięcznik poświęcony zagadnieniom pracy społecznej i opieki nad dzieckiem, p. 76–88.

TOMASZEWSKI, Jerzy/ŻBIKOWSKI, Andrzej (eds.), Żydzi w Polsce. dzieje i kultura. leksykon, Warszawa, Cyklady, 2001.

VOGELSTEIN, Hermann, Paula Ollendorff zum Gedächtnis, in: GSL P 28190 III, Jg. 15 (1938), Breslauer Jüdisches Gemeindeblatt. Amtliches Blatt der Synagogengemeinde zu Breslau, p. 1.

VRIES, Simon Philip De, Obrzędy i symbole Żydów, traduit par Andrzej BOROWSKI, Wznowienie., Kraków, WAM, 2001.

WEISSENBERG, Yvonne, Der Jüdische Frauenbund in Deutschland 1904–1939. Zur Konstruktion einer weiblichen jüdischen Kollektiv-Identität, Diss., Zürich, Universität Zürich, 2018.

WYSTOUCH, Boleslaw & ANTONIK, Roman Marjau, Kuryer Lwowski (Lemberger Courier), p. 6.

Debating Birth Control in Interwar Polish-Jewish Contexts
Ewa's Commitment to the Shaping of a Modern Jewish Polish Family Image

Heidi Hein-Kircher[1]

Abstract *In the second half of the 1920s, the changing understanding of the "new", independent, and self-determined woman led to extensive debates and a new understanding of the "family". An example of these interrelations that arose from early health feminist attitudes and the resulting self-empowerment in the interwar period is the Polish-Jewish weekly Ewa, published between 1928 and 1933. It reflected the nationalization of the "family", the negotiation of societal and individual attitudes towards birth control as well as that such attitudes were adapted in a particular way via a Polish-Zionist interpretation. The modern Jewish family should be healthy and prosperous, by being "cultured" and adopting a sophisticated approach to family planning. Ewa thus provided discussion of the challenges and tensions within a society which had only just begun to transform.*

The "black demonstrations" ("czarne protesty") that emerged against the introduction of a conservative law that almost banned abortion in Poland in 2016 have shown that women feel challenged and mobilized when they see threats to their reproductive behavior and health.[2] This example—among many in the

---

[1] Martin Opitz Library Herne – Affiliated to Ruhr University and Faculty for Historical Sciences at Ruhr University Bochum. This chapter is upon work from the project "Family Planning in East Central Europe from the Nineteenth Century until the Approval of the 'Pill'", funded by the German Ministry for Education and Research (BMBF, funding no. 01UC1902) and from COST Action 23149, supported by COST (European Cooperation in Science and Technology).

[2] Wiśniewska, Protest, 2017, p. 37–52.

Global North—illustrates the close link between women's (reproductive) health issues and the emerging "health feminism" movement fighting for related individual civil rights, because "health feminism" focusses particularly on problems regarding female sexuality and reproduction. In present times, we can observe particularly in extreme right wing populist discourses, which doubt these rights.[3]

However, this link has been observed since the turn from 19th to 20th century, when birth control became a politicized topic of intensive public discourse intertwined with moral reform and nationalization. In the age of nationalism, it became a means by which female activists could express their stance on the state of their nation and, at the same time, on the family.[4] Since then, birth control was debated not only among politicians but also by women who felt empowered to take a stand. Political debates about women's rights, which can also be interpreted as discourses of democratization, intensified especially in the interwar period. This took place against the backdrop of an increasingly changing attitude toward (female) sexuality and was promoted primarily by left-wing and liberal activists and feminists. In these discourses, sexuality was interpreted not only as a means of human reproduction, but also as a means of satisfying personal needs. Especially in the second half of the 1920s, the understanding of the "new", independent and self-determined woman, influenced by the complete liberalization of abortion in the Soviet Union and the economic crisis in Western Europe, led to extensive debates and, for example, mass demonstrations in the Weimar Republic demanding the liberalization of abortion.[5] The main issue was that women and couples should be able to decide for themselves the size of their families. Since then, "birth control" has become a mobilizing catchphrase for political demands for more female empowerment and social modernization and the use of birth control and the discourse surrounding it became a reflection of changing social (family) values and norms. This is particularly evident in the attitudes of socialist regimes toward birth control: on the one hand, they sought the complete liberation of women through socialism and therefore allowed abortions; on the other hand, there were also pro-natalist movements, for example in the People's Republic of Poland until the mid-1950s, which tried to make up for

---

3 Nicols, Women's Health, 2000, p. 56–64. DOI: 10.1111/j.1552-6909.2000.tb02756.x, see also: Weisman, Health Care, 1998.
4 See Articles in: Hein-Kircher, Hiemer, National Challenge, 2023, p. 3.
5 Usborne, Politics, 1992, Grossmann, Reforming Sex, 1995.

the population losses of the Second World War. Hence, a historical perspective on the discourses around reproductive behavior helps us to understand the agency of the main actors and the shaping of the image of the family. Such analysis contributes to a deeper understanding of social and value changes in societies[6] and, thus, of transforming family values and images.[7]

A very characteristic example of these interrelations that arose from early health feminist attitudes and the resulting self-empowerment in the interwar period is seen in the Polish-Jewish weekly *Ewa*, published between 1928 and 1933. This periodical reflected both the nationalization of the "family" and the negotiation of societal and individual attitudes towards birth control. It shows that attitudes were not just copied but adapted in a particular way via a Polish-Zionist interpretation. Thus, it clearly reflects the challenges of modern life within a socially and culturally changing community which was still under a strong religious influence.

For my argument, it is first necessary to provide context by examining the emerging Polish debates on birth control in the late 1920s; these debates have been topic of several scholarly publications but without a focus on customs and value changes.[8] Although *Ewa* has been explored in several smaller studies about the Polish-Jewish milieu,[9] there remains a gap with regard to its impact on shaping modern values and family images. Therefore, I discuss *Ewa*'s attitudes towards birth control based on a small number of representative articles in order to outline its contribution to value change and shifting family image.

The Emerging Debates on Birth Control in late-1920s Poland

By 1914, birth control was practiced by broad sections of the population,[10] while a fertility decline reflected a mental modernization of society, brought about by industrialization, urbanization and modern life.[11] The public negotiation of birth control and its individual and social consequences had been discussed intensively by politicians and activists. It was trans-nationally intertwined with

---

6   Hein-Kircher, Hiemer, Nešťaková, Norms, 2025.
7   Hein-Kircher, Werte- und Normenwandel, 2023, p. 60–74.
8   Latest study, with a social historical approach: Zielińska, Family 2022, p. 273–287.
9   For an overview of current research: Hein-Kircher, Werte- und Normenwandel, 2023. See also Hein-Kircher, National Challenge, 2023, p. 278–292.
10 Dienel, Kinderzahl, 1995, p. 55.
11 See Ciechanowski, "Is marriage so sacred?" in this volume.

eugenic and, later in the 1920s, "racial-hygienic" considerations through which reproductive behavior was becoming increasingly rationalized. Yet, as a result of rising nationalism, the "family" was seen as the basis of the nation and the nucleus of the state. The discourses that were triggered particularly by the radical German sexual reform-movement focused on abortion, which was for most women the only means of birth control, since only affluent women had access to modern contraceptives.

Although these trans-European discourses mooted birth control in a similar way, their specificities were acquired through the interweaving of national arguments. In Western Europe, abortion was already debated publicly in the decade before World War I. Under the conditions of partitioned Poland, a moral reform movement emerged, led by women activists. Their discourses were intertwined with abolitionist discourses against prostitution and the hygiene movement, in which the trafficking of (mostly Jewish) women and girls was seen as one of the most pressing moral problems. The moral reform movement contended with national destiny if marriage and family were not "pure" ("czyste") and viewed sobriety within marriage as the only means of birth control. Within this movement, Polish women felt mobilized to take a stance for their nation. This did not change until the late 1920s, when Jewish women also felt sufficiently empowered to advocate birth control.[12]

The main incentives came from outside the community. Across Europe, since 1918 the "new woman" had represented the dawn of democratic modernity in which traditional social orders such as marriage and, with it, the "duty to bear children", were being revised. The ideal of the "new woman" already represented social norms and value change, but particularly debates about birth control brought together the three key concepts of that epoch, "new", "freedom" and "sexuality". Hence, discourses on birth control brought mothers' freedom of choice—for whatever reason—increasingly into focus, not least since revolutionary Russia had legalized abortions.

In Poland, discourses on birth control and its synonym "conscious motherhood" ("świadome macierzyństwo") emerged only in the late 1920s and reflected the societal secularization and modernization of customs in a very specific way. In particular, during the six years from 1928 to 1934, debates on a new national law on marriage and abortion provided the basis for debates on "conscious motherhood", so that under the conditions of the authoritarian Sanacja

---

12   See Spielvogel, "The Jewish Women's League of Breslau" in this volume.

regime, the importance of birth control for Polish society were debated.[13] This was at a time when an economic crisis threatened society and anti-Semitism was on the rise. Only now, about 30 years later than in Western Europe, neo-Malthusian and eugenic arguments were woven into Polish debates beyond moral reform. As Elisa-Maria Hiemer points out in this volume,[14] the issue became a national one, as women were seen as having an important responsibility towards the new Polish state. Thus, birth control became a topic of strident political discourse because its acceptance and promotion were characterized as left-wing and liberal. The press, in particular, provided an arena in which it was part of intertwined discourses on women's emancipation, on the new marriage law and on sex education. Although birth control was intensively debated, in general Polish women's magazines did not devote much coverage and commentary to it.[15] The medical doctor and publicist Tadeusz Boy-Żeleński had fueled the debate with articles in *Wiadomości Literackie* (*Literary News*) and his feuilletons such as *Women's Hell*.[16] In this journalistic and ideological struggle, the Polish-Jewish weekly *Ewa* became a very specific resonance chamber for social and value change among Polish Jews, because of the extent to which debates about new family values and family planning were transculturally perceived and nationally adapted.

A Modern Stance Toward Jewish Family Conception

Ewa was published by Paulina Appenszlakowa between 1928 and 1933, at the height of Polish debates about birth control in the interwar period. Hence, it could be considered one of the main incentives for publishing the journal. *Ewa* referred to Tadeusz Boy-Żeleński's numerous publications but adapted their content to suit its own audience. It also frequently published opinions of expert activists like the gynecologist Herman Rubinraut. It is characteristic of the time that these and other men were presented as "experts" while women

---

13 For the harsh rejection of "conscious motherhood" by the Catholic press, See Marcin Wilk, "From Girls into Women, from Boys into Men" in this volume.
14 See Elisa-Maria Hiemer, "Divergent Narratives on Family Planning in Interwar Poland" in this volume.
15 Sierakowska, Elementy, 2004, p. 365–380.
16 Boy-Żeleński, Piekło kobiet, 1930. See Małgorzata Radkiewicz, "Single mothers and the issue of motherhood" in this volume.

writers were presented as activists and, in the case of Róża Melcerowa, a former Zionist deputy in the Sejm (1922–1927), as politicians.

Ewa reported extensively on the international women's movement in the USA, England and France, and especially in Weimar Germany, where the birth control movement reached its peak in the early 1930s.[17] In doing so, it held up mirrors to Jewish women and demonstrated to them the inherent force of such movements. For example, Melcerowa explained her attitude towards birth control by referring to her experience at the International Workers' Congress in 1926, quoting a German participant who asked her to tell their comrades in the East not to bring so many children into the world.[18]

Ewa devoted considerable space to contributing to the emergence of Zionist women.[19] It was particularly explicit about the connection between (Jewish) feminism and Zionism, combining secular, intellectual, emancipatory feminist and Zionist attitudes.[20] *Ewa* saw itself as an integral part of the developing Jewish nationalism[21] and was published primarily for middle-class and "cultured"[22] (i.e., modern and acculturated) Jewish women, who interacted with the publication through surveys and letters to the editor. It wanted to mobilize its audience for a "national and social-feminist emancipation".[23] The main editor Paulina Appenszlakowa's voice awarded the modern Jewish housewife a key influential role nationally, changing from the women's role of guardian of Jewish traditions such as *kashrut* and *nida* to custodians of the nation.[24] This highlights that the negotiation of new family values and norms, and thus of reproductive behavior, was combined with anti-religious instincts.[25] It reflected the specific experiences of modernity which challenged Jewish women to fulfill a conservative function as keepers of traditions and as mothers, while at the same time redefining their role in the face of fundamental general and also internal Jewish changes.[26]

---

17 See footnote 5.
18 Ewa, April 22, 1928.
19 Mickutė, Zionist Women, 2013, p. 151.
20 Leszczawski-Schwerk: Opieki społecznej, 2021, p. 396.
21 Plach, Feminism, 2005, p. 243.
22 Ewa, May 8, 1931.
23 Ewa, December 9, 1928.
24 Ewa, October 25, 1931.
25 Hein-Kircher, Werte- und Normenwandel, 2023, p. 69.
26 Steffen, Jüdische Polonität, 2004.

This tension was quite clear, e.g., in the assessment of pre-marital health examinations which were argued for on eugenic grounds. For example, *Ewa* regularly published pieces written by Melcerowa which echoed the discourses of the Weimar Republic, noting in November 1930 that pre-marital examinations were a "colossal benefit" both for the bride and groom and "for society as a whole".[27] One of her articles also pointed to the influence of eugenics, as she hoped that the battle against a "degenerate race" (as she characterized the poor, uneducated and "backward" Jews of the lower social strata) would be more successful through marriage certificates.[28] Like Melcerowa, Samuel Hirszhorn, a journalist working for the daily *Nasz Przegląd* (*Our Review*; edited by Appenszlakowa's husband),[29] considered birth control to be important for the Jewish nation, since he believed that the fewer children a family had, the more prosperous it would be.[30] Both authors therefore saw birth control as a tool for the renewal of the Jewish nation and the elevation of its culture to a modern one. Hirszhorn's opinion makes clear that discourses on birth control were rooted in and intertwined with those on societal modernization. By publishing such opinions, *Ewa* justified and linked attitudes towards birth control to a Zionist worldview.[31] Therefore, *Ewa*'s task first and foremost was to "take care of the education of mothers in order to awaken and to use their strength and energy [...] for the realization of the national, social and natural ideal".[32]

As she provided the quote in German, she implied what sort of educated audience she was addressing. Melcerowa thus contrasted the Eastern (i.e., Polish) living conditions of Jewish working-class families with those in Western Europe and ultimately provided a critique of the former.[33] She came to the conclusion that only children should be born if they were the explicit wish of their parents, because "no woman" wants a third, let alone a fourth or fifth child, and fewer children are loved all the more. Melcerowa thus combined modern ideas of the family with the Zionist need to create a "healthy" Jewish people. Overall, through such reports and columns, *Ewa* provided its audience with alternatives to the traditional understanding of the woman's role. We can there-

---

27 Ewa, November 16, 1930.
28 Ewa, November 16, 1930.
29 Similarly, this daily publication propagated an image of modern Jews: Szablowska-Zaremba, Wizerunek kobiety, 2016, p. 115–129.
30 Ewa, April 1, 1928.
31 Szabłowska-Zaremba, Portret, 2015, p. 545–560.
32 Ewa, December 9, 1928.
33 Ewa, April 22, 1928.

fore conclude that part of *Ewa*'s aims were to educate Jewish women according to an (imagined) ideal like that of (West) European women without deviating from the goal of making them Zionist women. According to Rafał Lemkin, one of the most distinguished jurists in the interwar period, the woman's "battle is double: for national and societal-feministic equalization".[34]

Under this premise, birth control was seen as the main tool for the making of a modern Jewish family. In an article in June 1928, *Ewa* asked (rhetorically) whether it was a national duty, and stated that birth control should be adapted to a family's economic, social and hygienic conditions. It also pointed out that Western societies were more advanced,[35] showing that they (especially Germany) were the point of reference and that birth control was a common practice in "all cultural nations".[36] *Ewa* thus concluded that having a large number of children was the result of a lack of (sex) education. This leads to the inference that *Ewa* not only saw sex education as important to avoid unwanted pregnancies, but also as a core part of "being cultured", since "a responsible cultured person must decide in favor of limiting offspring".[37]

Ewa's plea for sex education to prevent pregnancy was modified and combined with neo-Malthusian and eugenic arguments. One example of such an article was published by Melcerowa in April 1928. Since she supported Jewish orphanages and was familiar with poverty's consequences for Jewish family life, she argued that too many births threatened the health of the mother and of the new baby. Hence, she concluded that only as many children should be born as the parents wanted (and could afford). However, Melcerowa was aware that the majority of Jews in Poland lived in precarious conditions and that abortion by gynecologists could only be afforded by wealthy women. Therefore, "even the strictest religious prohibitions, threats and punishments from the legislature can prevent women from intervening in an unwanted pregnancy. [...] Such a heroic act can [ultimately] only be performed by a wealthy woman".[38] "Heroic act" implies a performance that goes beyond the individual and thus leads to the conclusion that Melcerowa also valued abortion as a deed that benefits the community. Her contribution thus exemplifies how activists-writers were well aware of the challenges and impact of the debates on birth control for Jewish

---

34   Ewa 4 March 1928.
35   Ewa 17 June 1928.
36   Ibid.
37   Ewa.
38   Ewa 22 April 1928.

women and could help to redefine Jewish women's consciousness with regard to the concepts of family and nation.

Conclusion

Contemporary debates on birth control reflected prevailing social and economic problems and processes, changing values in all national groups, and, as Katrin Steffen pointed out, difference too.[39] If Poland during the interwar period as a whole can be seen as a laboratory for the development of Jewish life in the modern era, the Polish-Jewish women's magazine *Ewa* contributed to it in a special way.[40] The example of *Ewa* shows that the discourses associated with family planning or the individual right to self-determined birth control were perceived transnationally. *Ewa* incorporated a Zionist orientation which stood up for bearing fewer but more healthy offspring for the benefit of the Jewish nation. The modern Jewish family should be healthy and prosperous, by being "cultured" and adopting a sophisticated approach to family planning. *Ewa* thus provided discussion of the challenges and tensions within a society which had only just begun to transform. Although *Ewa's* position was "Jewish-Zionist", its stance is ultimately analogous to contemporary debates in Christian societies: birth control was no longer understood as a private and (in the case of the Jewish family) a religious matter, but it concerned now the future well-being and strength of the nation. Although most of the arguments were similar in different societies, *Ewa's* were "tailored" to the respective needs of Jews and rendered appropriate for national (Zionist) identification. Consequently, the discourses on "conscious motherhood" could be characterized as the epitome of more general problems that also affected Jewish life in Poland. In line with Polish-Jewish attitudes that followed a specific interpretation and variation of Zionism, *Ewa* propagated the need to promote the prosperity and strength of the Jewish nation through "conscious motherhood". In doing so, it contributed to the emerging shape of the modern family image among the Polish Jewry of the interwar period.

---

39   Steffen, Polishness 2018.
40   Kassow, Jewish Street, 2014.

Bibliography

Appenszlak, Paulina/Wagmanowa, Iza (eds.), Ewa. pismo tygodniowe, Warszawa, P. Appenszlakowa i R. Iza Wagmanowa, 1928.

Dienel, Christiane, Kinderzahl und Staatsräson. Empfängnisverhütung und Bevölkerungspolitik in Deutschland und Frankreich bis 1918, Münster, Westfälisches Dampfboot, Theorie und Geschichte der bürgerlichen Gesellschaft, 1995.

Freeze, ChaeRan Y./Hyman, Paula/Polonsky, Antony (eds.), Jewish Women in Eastern Europe, Oxford; Portland, Oregon, The Littman Library of Jewish Civilization, (Polin. Studies in Polish Jewry, vol. 18), 2005.

Friedrich-Ebert-Stiftung, Forum Politik und Gesellschaft, Aufstehen für Frauenrechte, Berlin, Friedrich Ebert Stiftung, Gender matters!, 2017.

Grossmann, Atina, Reforming sex. the German movement for birth control and abortion reform, 1920 – 1950, New York and Oxford, Oxford Univ. Press, 1995.

Guzowski, Piotr/Kuklo, Cezary (eds.), Framing the Polish Family in the Past, London, Routledge, 2021.

Hein-Kircher, Heidi, Debating Social Change and the Jewish Nation. The Polish-Jewish Weekly Ewa on Jewish Families and Birth Control (1928–1933), in: Journal of Family History, 48 (2023), p. 278–292.

Hein-Kircher, Heidi & Hiemer, Elisa-Maria, Birth Control as a National Challenge. Nationalizing Concepts of Families in Eastern Europe, 1914–1939, in: Journal of Family History, 48 (2023), p. 235–244.

Hein-Kircher, Heidi/Hiemer, Elisa-Maria/Nešťaková, Denisa (eds.), Challenging Norms and Narrations. Family Planning and Social Change in Europe, New York, Oxford, Berghahn Books, 2025.

Hein-Kircher, Heidi & Nešťaková, Denisa, The Nation at Stake? Ideologizing Conceptions of Family Planning in East Central Europe Since 1939, in: Central Europe, 21 (2023), p. 68–77.

Hein-Kircher, Heidi, Ein Brennglas für Werte- und Normenwandel. Das Verständnis von Familienplanung von der Jahrhundertwende bis 1939 – Polen als Beispiel, in: Nordost-Archiv, 29. Begehren macht Akteur*innen. Praktiken der Subjektivierung im 20. Jahrhundert (2020), p. 60–74.

Janicka, Anna/Ławski, Jarosław/Olech, Barbara/Borkowska, Grażyna/Partyka, Jacek/Siedlecki, Michał (eds.), Żydzi Wschodniej Polski. Ser. 3. Kobieta żydowska, Białystok, Alter Studio, (Colloquia Orientalia Bialostocensia. Literatura/historia, vol. 13), 2015.

KASSOW, Samuel D, Oyf der yidisher gas, On the Jewish Street, 1918–1939, in: Polin. 1000 year history of Polish Jews, ed. Barbara KIRSHENBLATT-GIMBLETT, Warszawa, Museum of the History of the Polish Jews, 2014, p. 227–285.

KIRSHENBLATT-GIMBLETT, Barbara (ed.), Polin. 1000 year history of Polish Jews, Warszawa, Museum of the History of the Polish Jews, 2014.

LESZCZAWSKI-SCHWERK, Angelique, Między filarami opieki społecznej, pracy na polu kultury, upolitycznienia i feminizmu. Syjonistyczne "Koło Kobiet Żydowskich" we Lwowie (1908–1939), in: Studia Judaica, 48 (2021), p. 377–405.

MICKUTĖ, Jolanta, Making of the Zionist Woman. Zionist Discourse on the Jewish Woman's Body and Selfhood in Interwar Poland, in: East European Politics and Societies, 28 (2013), p. 137–162.

NICHOLS, Francine H., History of the Women's Health Movement in the 20th Century, in: Journal of Obstetric, Gynecologic & Neonatal Nursing, 29 (2000), p. 56–64.

PLACH, Eva, Feminism and Nationalism on the Pages of "Ewa. Tygodnik" 1928–1993, in: Jewish Women in Eastern Europe, ed. ChaeRan Y. FREEZE/Paula HYMAN/Antony POLONSKY, Oxford, Portland, Oregon, The Littman Library of Jewish Civilization, (Polin. Studies in Polish Jewry, vol. 18), 2005.

POLONSKY, Antony/WĘGRZYNEK, Hanna/ŻBIKOWSKI, Andrzej (eds.), New Directions in the History of the Jews in the Polish Lands, Warszawa, Academic Studies Press, 2018.

SIERAKOWSKA, Katarzyna, Elementy kobiecego dyskursu o seksualności na łamach międzywojennych periodyków dla kobiet, in: Kobieta i małżeństwo. społeczno-kulturowe aspekty seksualności. Wiek XIX i XX. Zbiór studiów, ed. Anna ŻARNOWSKA/Andrzej SZWARC, Warszawa, "DiG", (Kobieta, vol. 8), 2004, p. 365–380.

STEFFEN, Katrin, Contested Jewish Polishness. Language and Health as Markers for the Position of Jews in Polish Culture and Society in the Interwar Period., in: New Directions in the History of the Jews in the Polish Lands, ed. Antony POLONSKY, Hanna WĘGRZYNEK, Andrzej ŻBIKOWSKI, Warszawa, Academic Studies Press, 2018, p. 366–384.

STEFFEN, Katrin, Jüdische Polonität. Nation und Identität im Spiegel der polnischsprachigen jüdischen Presse 1918 – 1939, Göttingen, Vandenhoeck und Ruprecht, (Schriften des Simon-Dubnow-Instituts vol. 3), 2004.

SZABŁOWSKA-ZAREMBA, Monika, Portret syjonistki z łam "Ewy", tygodnika dla pań (1928–1933), in: Żydzi Wschodniej Polski. Ser. 3. Kobieta żydowska, ed.

Anna JANICKA, Jarosław ŁAWSKI, Barbara OLECH, Grażyna BORKOWSKA, Jacek PARTYKA, Michał SIEDLECKI, Białystok, Alter Studio, (Colloquia Orientalia Bialostocensia. Literatura/historia, vol. 13), 2015, p. 545–560.

SZABŁOWSKA-ZAREMBA, Monika, Wizerunek kobiety nowoczesnej na łamach "Naszego Przeglądu" 1923–1939, The portrait of a modern woman described in "Nasz Przegląd" 1923–1939, Prace Literaturozwnacze, 4 (2016), p. 115–129.

USBORNE, Cornelie, The Politics of the Body in Weimar Germany. Women's Reproductive Rights and Duties, London, Palgrave Macmillan UK, (Studies in Gender History), 1992.

WEISMAN, Carol S., Women's Health Care. Activist Traditions and Institutional Change, Baltimore, Johns Hopkins Univ. Press, 1998.

WIŚNIEWSKA, Agnieszka, Der Schwarze Protest hat Polen verändert, in: Gender matters. Infobrief zur geschlechterpolitischen Arbeit der Friedrich-Ebert-Stiftung, 7 (2017), p. 36–50.

ŻARNOWSKA, Anna/SZWARC, Andrzej (eds.), Kobieta i małżeństwo. społeczno-kulturowe aspekty seksualności. Wiek XIX i XX. Zbiór studiów, Warszawa, "DiG", (Kobieta, vol. 8), 2004.

ŻELEŃSKI, Tadeusz (Boy), Piekło kobiet, Warszawa, (Bibliotheka Boy'a), 1930.

ZIELIŃSKA, Agnieszka, The Jewish Family in the 19th and Early 20th Centuries, in: Framing the Polish Family in the Past, ed. Piotr GUZOWSKI, Cezary KUKLO, London, Routledge, 2022, p. 273–287.

"From Girls into Women, from Boys into Men"
An Expert's Discourse and the Press in a Medium-Sized City in Interwar Poland. The example of Tarnów

Marcin Wilk[1]

Abstract *This article answers the question of how expert discourses on sexuality, family planning, and marriage functioned in the local press of interwar Tarnów. By analyzing selected press titles from 1918–1939, the author explores the intersection of expert knowledge, gender norms, and religious influence in shaping public opinion. The study sheds light on the role of medium-sized cities in the circulation of modern ideas in interwar Poland.*

Introduction

The study of diverse historical contexts pertaining to sexuality, family planning, and reproduction yields insights into the socio-cultural and political shifts within the regions and communities under consideration. In Europe, the advent of the twentieth century heralds a distinct epoch in these transformations, particularly concerning the youth, who served as pivotal actors in the transition. In Poland, as in many other Central and Eastern European countries, the processes of modernization and emancipation accelerated after 1918. New knowledge about human beings, their gender and sexuality, was not only a result of scientific discoveries, however. Attitudes towards sexual behaviour and expectations towards e.g. "female youth" and "male youth" were regulated by the voice of public opinion too. So far, research on this topic has focused on identifying certain general trends in sex education, ethnic discourses, or

---

1 Institute of History, Polish Academy of Sciences.

knowledge transfers and the formation of expertise.[2] Studies have brought a picture of general trends. The discourses of micro-regions or smaller – than Warsaw, Krakow or Lviv – provincial centers were much less often analyzed. Meanwhile, relatively well available sources on the subject, primarily the press, offered one of the basic tools for expressing opinions in society in a medium-sized city. Apart from local reports, one found opinions on old and new social ideas and how they were practiced. In this way, the voice of so-called experts – people recognized as such in a specific field – spread to the local community. The study of both expert functioning and expert discourses, therefore, expands not only the knowledge of social mechanisms, but opens new perspectives for regional and smaller city studies, which were often important centers of knowledge flow, cultural practices and new social norms in interwar Poland.

In this paper, I will focus on interwar Tarnów, a medium-sized Polish city located in the former western Galicia, about 100 kilometers east of Kraków, and inhabited then by a Polish-Jewish community. Between 1918 and 1939 the town metamorphosed significantly.[3] This article focuses on the study of expert discourses in the press of interwar Tarnów. I define expert discourse here quite broadly – such parts of speech that refer to or contain statements of expert type, that is, refer, directly or indirectly, to – as Foucault argues – 'purified and neutral' scientific language.[4] As such they also have a normalizing character. The source basis is eleven press titles published in Tarnów between 1918 and 1939 concerning the city's issues.[5] In the research sample there were no magazines representing Jewish and Zionist communities[6] – access to them turned out to be limited in the Covid era. At the same time the very present role of the Catholic Church is noticeable. The influence of the Catholic Church on the pro-

---

2   Gawin, Sex reform, 2008, p. 181–186. Mickute, Modern, 2011. Karczewski, Homosexuality, 2022, p. 1–18. Kościańska, Gender, 2021.
3   Inhabitants: 35 347 (1921), 55 642 (1938). Gołębiowski, Stosunki, 1938, p. 414.
4   Foucault, Histoire, 1976.
5   Weekly magazines: Goniec (1934), Hasło (1926–1939), Lud Polski (1919–1926), Nasz Głos (1925–1929), Nasza Sprawa (1933–1939), Praca (1924–1935), Słowo Tarnowskie (1927–1929), Wiadomości Tarnowskie (1933–1934), Ziemia Tarnowska (1938–1939), and the monthlies magazines: Świt (1934–1938) and Własnemi Siłami (1936–1939). The research covered selected years.
6   On Jewish and Zionist positions of the debate see Hein-Kircher, "Debating Birth Control in Interwar Polish-Jewish Contexts" and Izabela Spielvogel, "The Jewish Women's League of Breslau" in this volume.

duction of the press in the city was strongest in the second half of the 1930s.[7] I use the press as one possible tool of distribution, and in my analysis, I have used a variety of genres – from broadsheet journalism, to reprints of official or advice texts, to advertising. I consider the press as an important tool in the processes of the democratization of urban life and the formation of public opinion. The leading recipient of the press was the intelligentsia, however the ambitions of many editors, who tried to address the peasant or worker class of the "Polish folk", should be emphasized as well.

Although the press has been used quite often in studies of smaller urban centers, it has rarely used as material for the study of expert discourses or the circulation of knowledge about sexuality. This article, predominantly employing gender history methodologies,[8] endeavors to address this gap, at least partially.

The main purpose of this presentation is to identify traces of expert discourses in press narratives in Tarnów. I am interested in answering the questions: Who was the expert for public opinion in interwar Poland? What parts of expert discourse permeated the press? What conditioned their audibility? Among many issues, I focused on the issues related to the discussion on marriage, as well as on the issues of motherhood and sexuality. An important aspect of the research presented here is the gender history. The issue of socialization was also a key context in my research.

Social Factors of Expert Knowledge in the Interwar City

Before I discuss the examples, it is worth starting with the question of what an expert was and explore his/her social role in the examined period. In the dictionary of Polish language by Michał Arcta (1916) an expert is someone "proficient in something, has chosen to study it; a knower, appraiser; someone knowledgeable, professional, a specialist".[9] Sociologist Florian Znaniecki in his essay "Uczeni polscy a życie polskie" ["Polish Scholars and Polish Life"] (1936), in turn, specified that an expert is a person educated and prepared by a learned professor to "cope with the life situation" in the academic system, related to scientific

---

7   Lachendro, Prasa, 2006, p. 305 – 326.
8   Scott, Gender, 1986, p. 1053–1075.
9   Arcta, Słownik, 1916, p. 286.

research.[10] In interwar Tarnów, the number of experts who specialized in medical matters was limited to 40 medical doctors working in the city, including three railway doctors, six dentists, and six dental technicians and midwives. There was also a general hospital, a Garrison infirmary, a children's clinic, a home for incurables, a district dealth fund and six pharmacies.[11] The newly opened prison also had a hospital. After Poland regained independence in 1918, the tasks of implementing sanitary regulations, controlling doctors, surgeons, and pharmacists, as well as dealing with healers and charlatans, and fighting infectious diseases in humans and animals were taken over by the district doctor and the city doctor. All of them were social workers, although it should be noted that access to health care largely depended on the individual means of citizens, so in many cases citizens, especially the poorer ones, remained without genuine health care. It is worth mentioning that many of the doctors, almost all of them Jews, who worked in the county health centre (opened in 1930) were very poorly paid.[12] In mentioning the numbers, two disclaimers must be made: firstly, the figure does not include the many professionals that applied medical knowledge in their work, such as educators or hygienists. Secondly, expert discourses may have been used in public debates by non-specialists, such as representatives of religious organizations.

Although among the sanitary councillors there were no women on the management staff, it is worth indicating here that the monthly magazine "Własnemi siłami" ["By Our Own Efforts"] was edited by female Catholics, and Marianna Dmochowska, the editor and the author of many texts in the magazine. She was also the general secretary of the Catholic Women's Association. Although the section "Dbaj o zdrowie" ["Take care of your health"] published in "Własnemi siłami" is unique,[13] and the focus is on dental hygiene, there is no mention of venereal diseases or other manifestations or consequences of the sexual life among the citizens. Expert knowledge was also rather rarely invoked in discussions about young people.[14]

A characteristic feature of the social innerworkings of knowledge in interwar Tarnow – as shown by reading the newspapers – was its competitiveness with non-expert knowledge, especially that based on the dogmas of faith. For

---

10   Znaniecki, Społeczne role, 1984, p. 229.
11   Maniak, Ochrona zdrowia, 1938, p. 705.
12   Ibid. p. 707.
13   Dbaj o zdrowie, 1937, 1938.
14   Wilk, Prasy lokalnej, 2021.

example, in 1926, the Catholic "Nasz Głos" ["Our Voice"] reprinted a speech by Tarnów bishop Wałęga from 1911, who explicitly stated that "knowledge and science are treated as God when God's revelations are ignored".[15] For this reason, experts probably had to deal with distrust in the city. In one article, a doctor encourages people to behave during a visit as if they were going to confession and thus to be honest about what may have affected the patient's condition.[16]

Finally, despite the fact that the present research focuses on the press published in Tarnów, it is worth remembering that regional, national, and even foreign newspapers reached and were read in the city. It was distributed through private channels, but was also available in schools or local reading rooms, and even in the local Warszawianka café.[17]

Marriage and its Crises

One interesting theme indirectly related to expert discourses concerns the transformation of the institution of marriage. The roles in marriage – different for girls/women and boys/men – were determined by the gender binary. Echoes of representations of traditional roles permeated the law. As Claudia Kraft has noted, this emphasized the bipolarity of the structure of the organic world, consisting of a male element and a female element, "endowed in the human world with specific physical and psychological attributes".[18]

Remarkably, more emphasis on the future role of wife or mother was imposed upon girls. They were the ones from whom more was demanded, thus they were presented with a more extensive catalogue of duties. They had to prepare not only themselves, but also to be, as it were, for difficulties on the part of the man and marriage as such. In the article "For female candidates for marriage" published in "Praca" ["Labour"], for example, we read: "A woman's happi-

---

15 Wyjątki z mowy JE. bs. biskupa Wałęgi na Kongresie Marjańskim w 1911 r. w Przemyślu, "Nasz Głos" 20 (1926), 4.
16 Markowski, Byłam u lekarza, 1936, p. 4.
17 [advertisement], Ziemia Tarnowska, 1938, p. 6. Sprawozdanie roczne z działalności szkoły za rok szkolny 1923/1924, "Zespół Szkół Ekonomiczno-Gastronomicznych w Tarnowie", State Archive in Tarnów (hereinafter: SAT) 33/752/244, 11; Sprawozdanie roczne z działalności szkoły za rok szkolny 1923/1924, "Zespół Szkół Ekonomiczno-Gastronomicznych w Tarnowie", SAT 33/752/244, 48.
18 Kraft, Równość i nierówności, 2004, p. 317. More about marriage, See: Pietrzak, Sytuacja prawna 2000, p. 82. See also Ciechanowski, Is marriage so sacred? in this volume.

ness in marriage consists of understanding her husband and being indulgent towards his weaknesses or habits. A lot of disappointments would be avoided if girls wanted to understand that their husbands are made of the same material as their fathers or employers [...]".[19] Interestingly, the vision of the man's role in this perspective seems to depend on the fulfillment of the wife's conditions. As the journalist says, she shouldn't persecute her husband with constant excuses and suspicions, but should rather be *"a good sport"*.[20]

Marriage as a burden, duty, anguish – such images often appeared in various narratives, but in Catholic journalism they gained the dimension of a distant promise. Fulfilling the role of a wife – that is, sometimes an anguished one – meant happiness and empowerment, even if it was paid for with "long hours of waiting for your husband, for all those bad moments of loneliness that bleed your heart so much, when your husband's house will no longer be occupied, when he will be impatient and rough, when he will not appreciate your work, the immensity of your love, when he will look for a quarrel in your words" – as Janina Pietrusińska enumerated in the pages of "Własnemi siłami".[21] The final advice, in accordance with the requirement of durability of marriage, was, of course, to persevere, despite the torments, although there is an interesting thread of interest in the psychological states of women, as well as in the indifference of the husband to the affairs of the home. Single women were also recognized.[22] The psychological price that a woman has to pay to fulfill the role of mother, wife, or grandmother expected of her often appears in Catholic narratives. The duty of parenting in the home rested on the woman. The press avoided commenting on the details, as if the issues of caring for a young child related only to the private sphere and concerned matters not dealt with by the public. Catholic narratives rarely defined concepts explicitly, rendering issues taboo that related to corporality or the physical side of the conjugal act. At the same time, sexual abstinence was required of spouses. In a pastoral letter reprinted by "Nasza Sprawa" ["Our Business"], Bishop Franciszek Lisowski reminds that marriage is "a sacrament of the living, that is the newlyweds should have pure souls".[23]

---

19   Dla kandydatek do małżeństwa, 1933, p. 4.
20   Ibid.
21   Pietrusińska, Źródło, 1937, p. 4.
22   Własnemi siłami, 1939, p. 8; 1939, p. 8.
23   Lisowski, List pasterski, 1934, p. 5.

In the late 1920s and early 1930s the process of law unification in Poland, and more specifically the proceedings on the Marriage Act, brought a certain impetus to the debate on the status of marriage and its functioning in society and the state. The fundamental desire of the Codification Commission was that the existing Code should reflect "the spirit of the requirements of modern civilization".[24] At issue were, among other things, divorces. The Catholic Church treated marriage as a sacrament, an indissoluble bond, outside the law. Hence it did not agree to divorce. Moreover, it criticized the moral awkwardness of the state, referring to the Constitution, but also to the privileged – regulated by the Concordat – position of the Catholic Church.[25] However, the transformation of marriage was a fact. The conventionality of marriage was increasingly recognized. An interesting text on the subject is provided by the "Goniec" ["Messenger"], which calls marriage an "institution" and emphasizes the deterioration of labour markets, which prevents men in particular from establishing their own households. The text introduces the concept of the "mate marriage". "Perhaps what we call today heresy, utopia, backwardness and even ‚immorality' will in a few years' time be a normal phenomenon, sanctioned by law".[26] It is also accepted that the process of change within the institution came from the woman. This text is in fact unique, because divorce was rarely been written about in the press. Generally, when writing about difficult matters, where experts already agreed, but where customs held back – they used examples from distant countries or cultures. This is a recurring narrative motif, and it seems that such a strategy made it possible to become accustomed to change. In the studied press we find several extensive texts on this subject. For example, divorce is written about in relation to ancient Rome.[27] In this context new media, such as cinema and radio, seems to be a good tool for spreading expert knowledge. Celebrity life normalized the transformation of the institution of marriage. This is a helpful lens for interpreting reports about Pola Negri's subsequent marriage, for example.[28]

---

24 Kraft, Równość i nierówności, 2004, p. 312.
25 Prawa małżeńskiego, 1934, p. 13.
26 Małżeństwo na próbę, 1934, p. 6.
27 Rozwody, 1929, p. 3. Przymus żenienia, 1923, p. 10.
28 Czwarte małżeństwo Poli Negri, 1934, p. 4.

Motherhood

As already mentioned, the Catholic narrative argued that a woman's natural vocation was motherhood. It claimed a woman was meant to take care of her child from its very conception, she should conduct herself "soberly and morally" (Bishop Lisowski). Long texts on this topic were published by Catholic magazines, especially after the discussion on marriage. Pastoral epistles were reprinted there. In one of them, addressed to believers from all over Poland, we read about "conscious motherhood" in the "Catholic spirit", which is defined here as "a deep sense of the dignity of the mother and her duties both in terms of health and hygiene, and above all in terms of raising the offspring. [...] On the other hand, we must condemn intentional motherhood understood and practised as the prevention of births by unlawful means. Under the name of 'conscious motherhood', certain groups are propagating the prevention of motherhood and childbirth. This movement should properly be called the 'conscious maternity' movement under the slogan 'fewer births'. Conscious motherhood understood in this way is a consequence of the materialistic understanding of the family, which is also evidenced by the reference made by the Polish conscious motherhood movement to such a movement in the Soviet countries. This movement is wrongly justified through slogans of hygiene, social justice, correction of nature, and even the good of the State. In fact, it is the desire for the law of sin".[29] The multi-layered rhetorical construction of this statement is interesting for many reasons, although a few aspects are worth highlighting. Firstly, the association of "modern culture" was often associated – in the perspective of the Catholic narrative – with "conscious motherhood". Secondly, an attempt to regain, in a way, the expression "conscious motherhood", which – according to some of the women's discourses of the twentieth century in Poland – was most often associated with the separation – at least partial – of procreation from sexual pleasure. The bishops' letter, addressed to the whole of Poland, was reprinted in "Nasza Rzecz" ["Our Thing"], a Catholic weekly with a high circulation for the city. It is not known what audience it reached or how wide an impact it had.

Catholic journalism was critical of "lectures, counselling centres and brochures", and did not approve of such initiatives. In the perception of

---

29 Wspólny List pasterski, 1934, p. 2. On the criticized position towards 'conscious motherhood' See Hein-Kircher, Debating Birth Control in Interwar Polish-Jewish Contexts and Radkiewicz, Single mothers and the issue of motherhood in this volume.

Catholic publicists, they were associated with "pornography", which was particularly dangerous for the youth. In reality, in secondary schools in Tarnów, for example, there were lectures on venereal diseases held among 3rd grade students, but these were intended only for boys.[30] In fact, courses and papers – for example, on the care of mother and child read at meetings of the Women's Civic Labour Union[31] – were often addressed to the adults.

Sexuality Issues

The Tarnów press devoted relatively little space to sexuality. The lack of appropriate vocabulary[32] made it almost impossible to talk about issues related to corporeality, physiognomy, or sexuality directly. They had to resort to satire and allusions. However, many topics were above all tabooed. In addition to indirect allusions, there was also information in the press about, for example, the women's trade or "streetwalkers". For example, I have not been able to find extensive texts on the epidemic of venereal diseases, which, as we know, posed a social problem especially in the 1930s.[33] In the capital, an anti-venereal law was being drafted, which was supposed to regulate matters of health care by imposing a standard number of beds reserved for venereal patients in hospitals. This obligation involved cities with populations over 25,000, and that included Tarnów.

While the topic of sexuality was not dealt with directly, I would like to draw attention to two interesting texts, which can provide contextual guidance to the issues discussed here. The first concerns dance, and its modern, fashionable form: "Today's dances, or so-called fashionable dances, almost without exception express the excitement of lovemaking. The man seeks, the girl or woman agrees, and then they both begin to rejoice in their mutual closeness and momentary possession of one another". Interestingly, the text quotes experts and specialists, including Havelock Ellis, a British physician and founder of modern sexology[34]. The second text is a reprint published in "Nasz Głos"

---

30   Sprawozdanie roczne z działalności szkoły za rok szkolny 1927/1928, "Zespół Szkół Ekonomiczno-Gastronomicznych w Tarnowie", SAT 33/752/248, 15.
31   Hasło, 1933, p. 2.
32   Sierakowska, Elementy, 2004, p. 369.
33   See: https://dlibra.umcs.lublin.pl/dlibra/plain-content?id=10086 (05.06.2025); See also: Grata, Walka, 2013, p. 253–274.
34   Na karnawał, Praca 7–8 (1933), p. 2.

("Our Voice"). The text – "Flapper" – refers to the Western model of a new girl, a representation of gender in the 1920s.[35]

Finally, it is also important to notice the absence of expert discourses in the secondary school magazines. Sexuality in the official circulation was strongly tabooed.[36]

Conclusions

The analysis of Tarnów's press texts shows that expert discourses found it rather difficult to penetrate journalistic discourse. This transfer was limited by the lack of language and the tabooization of issues directly related to sexuality, the body or eroticism. Those who published used various rhetorical strategies – allusions or references to distant examples. At the same time, the identified traces of expert discourses from the press reflect only certain ideas about marriage or sexuality, and only to a small extent do they reflect their actual image. It is much more common for male experts to speak in the press, although the influence of women is becoming stronger. The analysis of press releases shows that Tarnów was getting into the swing of processes of a more universal character, such as the emergence of new personal patterns within gender roles, such as the "the modern girl!".[37] However, the local press used here as a source base does not give a full picture of the complexity of this phenomenon. In order to gain a deeper insight into this phenomenon, it would certainly be necessary to make use of other types of material, such as diaries or prosecution material – and above all, material that completes the bi-ethnic character of the city: the press or Jewish documentation.

It is worth noting in the context of the research results presented here that the specificity of medium-sized centers is often defined by the role these cities play in constituting a transfer tool on the path of upward social mobility from smaller to larger centers. It would be interesting to compare the results of the research conducted in Tarnow with other centers of this type both within Poland and in other Central and Eastern European countries. How did the expert discourses differ? What determined their audibility? What was conducive

---

35   Chłopczyca, Nasz Głos 32 (1927), p. 3.
36   One of the few exceptions is the story of the first date written by a girl in the monthly school youth magazine "Świt": Moja pierwsza randka, Świt 4 (1935), p. 10.
37   Nicholas, Modern Girl, 2015.

to expanding the field of expert discourse? What impact did popular culture, especially cinema, which was a medium for new ideas and new definitions of marriage or reproduction? These questions above all in relation to the issue of sexuality or family planning seem relevant and interesting. Comparative research can provide new knowledge on the social and cultural contexts of the historical changes of the interwar period.

Bibliography

ARCT, Michał, Słownik ilustrowany języka polskiego. T. 1, (A-O)., Warszawa, M. Arct, 1916.

FOUCAULT, Michel, Histoire de la sexualité, 1. La volonté de savoir, Paris, Gallimard, 1976.

GAWIN, Magdalena, The Sex Reform Movement and Eugenics in Interwar Poland, in: Studies in History and Philosophy of Biological and Biomedical Sciences 39 (2008), p. 181–186.

GRATA, Paweł, Walka z nierządem w polityce państwa polskiego w latach 1918–1939, in: Margines społeczny Drugiej Rzeczypospolitej / red. nauk. Mateusz Rodak., ed. Mateusz RODAK, Warszawa, Instytut Historii PAN, (Metamorfozy Społeczne, vol. 6), 2013, p. 253–274.

KARCZEWSKI, Kamil, Transnational Flows of Knowledge and the Legalisation of Homosexuality in Interwar Poland, in: Contemporary European History, (2022), p. 1–18.

KIRYK, Feliks/RUTA, Zygmunt (eds.), Tarnów. Dzieje miasta i regionu. [T.] 2. Czasy rozbiorów i Drugiej Rzeczypospolitej, Tarnów, Urząd Miejski w Tarnowie, 1983.

KOŚCIAŃSKA, Agnieszka, Gender, Pleasure, and Violence. The Construction of Expert Knowledge of Sexuality in Poland, Bloomington, Indiana University Press, 2021.

KRAFT, Claudia, Równość i nierówności w II Rzeczypospolitej. Prawo małżeńskie w dyskursie publicznym na przełomie lat dwudziestych i trzydziestych, in: Kobieta i małżeństwo. społeczno-kulturowe aspekty seksualności. wiek XIX i XX. zbiór studiów, ed. Anna ŻARNOWSKA, Andrzej SZWARC, Warszawa, DiG, 2004, p. 311–327.

LACHENDRO, Jacek, Prasa województwa krakowskiego w latach 1918–1939, Jacek Lachendro., Kraków, Towarzystwo Wydawnicze "Historia Iagellonica, Studia z Historii XX Wieku"; vol. 4, 2006.

MICKUTÉ, Jolanta, Modern, Jewish, and Female. The Politics of Culture, Ethnicity, and Sexuality in Interwar Poland, 1918–1939, Ann Arbor, Mich., Proquest, 2011.

NICHOLAS, Jane, The Modern Girl. Feminine Modernities, the Body, and Commodities in the 1920s, University of Toronto Press, 2015.

PIETRZAK, Michał, Sytuacja prawna kobiet w Drugiej Rzeczypospolitej, in: Równe prawa i nierówne szanse. kobiety w Polsce międzywojennej. zbiór studiów, ed. Anna ŻARNOWSKA/Andrzej SZWARC, Warszawa, DiG, 2000, p. 77–91.

RODAK, Mateusz (ed.), Margines społeczny Drugiej Rzeczypospolitej / red. nauk. Mateusz Rodak., Warszawa, Instytut Historii PAN, (Metamorfozy Społeczne, vol. 6), 2013.

SCOTT, Joan W., Gender. A Useful Category of Historical Analysis, in: The American Historical Review, 91 (1986), p. 1053–1075.

SIERAKOWSKA, Katarzyna, Elementy kobiecego dyskursu o seksualności na łamach międzywojennych periodyków dla kobiet, in: Kobieta i małżeństwo. społeczno-kulturowe aspekty seksualności. wiek XIX i XX. zbiór studiów, ed. Anna ŻARNOWSKA, Andrzej SZWARC, Warszawa, DiG, 2004, p. 365–380.

WILK, Marcin, Młodzież międzywojennego miasta w świetle prasy lokalnej. Przypadek Tarnowa, in: Roczniki dziejów społecznych i gospodarczych, 82 (2021), p. 183–215.

ŻARNOWSKA, Anna/SZWARC, Andrzej (eds.), Kobieta i małżeństwo. społeczno-kulturowe aspekty seksualności. Wiek XIX i XX. Zbiór studiów, Warszawa, DiG, 2004.

ŻARNOWSKA, Anna/SZWARC, Andrzej (eds.), Równe prawa i nierówne szanse. kobiety w Polsce międzywojennej. Zbiór studiów, Warszawa, DiG, 2000.

ZNANIECKI, Florian, Społeczne role uczonych, Warszawa, Państwowe Wydawnictwo Naukowe, 1984.

Divergent Narratives on Family Planning in Interwar Poland
Between "secret marriage tricks" and "the obligation of maternity"

Elisa-Maria Hiemer[1]

> "The house becomes a hell, the wife in her thirties is ill, an old woman, physically repulsive; the husband, stunned by the children's cries, seeing his wife either pregnant or nursing, or both together, runs away from home to the pub, where his drinking makes the family's misery even worse."[2]

Abstract *This article explores interwar narratives on birth control across political, educational, and private spheres, revealing conflicts despite public liberalization. It examines expert discourse on unwanted pregnancies, varying educational approaches, and women's real-life experiences in abortion trials. Highlighting micro-historical sources, it underscores the challenges of singular proof and the necessity of contextual analysis.*

This quote from the writer and women's rights activist Tadeusz Boy-Żeleński not only illustrates roles and gender images in families, it also sheds light on the misery of families with many children in interwar Poland. The country

---

1   Freie Universität Berlin, Friedrich-Meinecke-Institut.
2   Boy-Żeleński, Piekło kobiet, 1930. https://wolnelektury.pl/katalog/lektura/pieklo-kobiet.html (05.06.2025).

has been reestablished after 123 years of partition between the Prussian, Habsburg, and Russian empires. The family as a sociodemographic factor was crucial to political and social discussions about Poland's future. In parallel, women's rights activism emerged around 1900 and criticized both the "motherhood mandate"[3] and the social and health risks women with children were exposed to and challenged preexisting ideas on female sexuality and unwanted pregnancies.[4] The "moral reform" with its liberal concepts of motherhood and partnership provoked lively debate in the media and literature, among law and medical experts. Prior to the codification of Polish law in 1932 and its new liberal regulations on abortion, the debates on birth control and family planning had developed vociferously.[5] New and improved forms of relationships (e.g., dating, cohabitation and "amical marriage")[6] became more popular and even vending machines with 'hygienic gums' being made available in bigger cities.[7] These developments ushered in an era where sexuality without reproductive goals became more accepted and detached from the field of adultery – at least gradually.[8] Nonetheless, the Polish discourse was highly diverse even in a milieu of similar ideological beliefs.[9]

Social history is often told as a history of progress as the anthropologist and specialist for the history of sexuality, Agnieszka Kościańska, pointed out.[10] Based on this observation, the paper scrutinizes the divergent narratives in three overlapping domains: political, educational, and private.[11] The process of diversification of interpersonal relations in the interwar period has become a core topic of contemporary research,[12] and since the 2000s, an increase in

---

3   Russo, Mandate, 1976, p. 143–153.
4   Kałwa, Kobieta aktywna, 2002.
5   Krajewska, Spór, 2013, p. 275–300.
6   Janicki, Hipokryzji, 2015, p. 80.
7   Nonetheless, the fact that condoms would be available day and night even to minors made authorities handle their distribution in a restrictive manner. Archiwum Akt Nowych (AAN): Ministerstwo Opieki Społecznej w Warszawie, f. 1582, 72.
8   Gawin, Sex Education, 2009, p. 217–235.
9   Krajewska, Spór, 2013, p. 275–300.
10 Kościańska, Moose 2021. Kościańska, Pleasure, 2021. Kuźma-Markowska, antykoncepcji, 2009, p. 603–619; Kuźma-Markowska, Drop of Milk, 2011, p. 131–147.
11 Research for this article has been conducted in the project Familienplanung im östlichen Europa vom 19. Jahrhundert bis zur Einführung der "Pille" funded by the Federal Ministry of Education and Science BMBF 01UC1902.
12 Landau-Czajka, Procesy socjalizacji, 2013. Żarnowski, Państwo i społeczeństwo, 2014. Żarnowski und Mędrzecki, Metamorfozy społeczne, 2015.

research on female sexuality and agency[13] has been noticeable. However, the topic of abortion remains discussed on public levels like politics or intellectual and educational movements only.[14] Approaches on microhistorical level[15] are rare. With a short analysis of the testimonies given during an abortion trial held 1928–1929 at the Suwałki county court, this article wishes to enhance knowledge about the individual's life reality.[16]

It shows that the process of liberalization in the public sphere only had a very limited impact on the personal sphere and especially to women living in precarity. To make this divergence visible, I will first shed light on the experts' discourse on unwanted pregnancies. Secondly, I will outline divergent educational approaches in advice literature and finally, I will contrast these findings with the real-life experiences of women as they are presented in trials on abortion. This multi-layered approach intends to raise awareness regarding the character of micro-historical sources in research. Though historical in scope, this text tackles a timely issue: the topic of female self-determination and political decision-making experiences a global backlash nowadays.[17] Abortion is still not considered as means of health care in most of the countries. With Poland being the worst ranked country for years according to the *Contraception Policies Atlas* published by European Parliamentary Forum on Sexual and Reproductive Rights, many countries do not prioritize reproductive health care by reducing the barriers to accessing information, procedures, or reimbursement.[18] Abortion still carries a stigma for women who terminate their pregnancies: in the interwar period as well as today.

---

13 Żarnowska und Szwarc, Równe prawa, 2000. Sierakowska,. Elementy kobiecego dyskursu, 2004, p. 365–380.
14 Slany, Regulacja prawna aborcji, 2006, p. 139–144. Szlagowska, Świadomym macierzyństwem, 2018, p. 31–47.
15 Like this article shows for a 1949 case: Klich-Kluczewska, Przypadek Marii, 2012, p. 195–209.
16 Hiemer, Maternity, forthcoming.
17 Beláňová, Anti-abortion activism, 2020, p. 395–413. Ziegler, Roe v. Wade, 2022, p. 16–21.
18 European Parliamentary Forum for Sexual and Reproductive Rights, Policy Atlas, 2023. https://www.epfweb.org/sites/default/files/2023-02/Contraception_Policy_Atlas_Europe2023.pdf (05.06.2025).

The Experts' Discourse. Rethinking the *Matka Polka* Topos

The founder and prominent representative of forensic medicine in the Second Polish Republic (1918–1939) was Wiktor Grzywo-Dąbrowski. He argued that there was hardly any other topic in law that was so opposed to society's view as the punishment of abortions. He blamed the patriotic view on motherhood for this problem, and especially the topos of *Matka Polka*, the Polish mother as the bearer of a nation that had just emerged after 123 years of partition.[19] The public moral verdict condemned casual sexual behaviour and often considered an unwanted pregnancy to be the woman's fault since "she showed a strong desire to have relations with men",[20] as stated in some investigative police material. Indeed, it was harder for a woman who miscarried a foetus out of wedlock to defend her innocence. Like other Polish jurists, Grzywo-Dąbrowski called for a fundamental change in the perception of single mothers. For him, the main aim was to prevent abortions, for example by allowances for non-married women to raise a child born of non-normative relationships.[21] Whereas in these cases, experts mainly had the same point of view, the topic of abortion remained controversial and the discourses about it showed an inherent paternalist view on the women's body: Even liberal voices, which acknowledged poor living conditions and insufficient social security as main factors for the high rate of illegal terminations,[22] acknowledged that there was more than a woman's personal interest to be considered. The lawyer Stefan Glaser identified three groups of interest around which the discussion was centered: the fetus, the state, and third parties (including the woman).[23] However, he concluded that "the state cannot force a woman to carry a child to term and to give birth to it."[24] With this statement, he clearly opposed German opinions of that time, e.g.: "if a woman consents to intercourse, she signs a

---

19   For the role of Polish women during the partition period and their emancipation see e.g. Walczewska, Feministki, 1999. Staśkiewicz, Polin, 2018, p. 109–117.
20   Archiwum Państwowe w Katowicach: Policja Województwa Śląskiego, f. 290, 46.
21   Grzywo-Dąbrowski, Przerwanie ciąży, 1926, p. 31.
22   The numbers for the interwar period vary. The Statistical Yearbook of Poland counts between 1233 to 1399 registered cases of illegal abortions in the period 1924–1928. Since the procedure was mainly performed clandestinely, estimates assume up to half a million abortions between 1922 and 1938. See Łodyga, Metody, 2016, p. 244.
23   Glaser, Kilka uwag, 1927, p. 39–40.
24   Ibid.

contract with the State to give children to it."²⁵ I argue that questioning the legal regulations of the former partition powers was a crucial part of Poland's legal emancipation. Glaser was in favor of the woman's freedom of choice as it assured her physical integrity. But this approach was rather exceptional since the majority of (exclusively male) experts did not ascribe agency to the woman. Grzywo-Dąbrowski for example required the family's consent to surgery, i.e. the husband's final decision.²⁶ Additionally, the surgery itself remained reprehensible in moral terms. A member of the Codification Commission, Karol Czałczyński, concluded on the judgement of abortions: "Killing a fetus is a crime. However, we must realize that in this case we are dealing with the intrusion of state power into the domain of the confidential existence of the individual. (...) However, the specific social, economic, and legal conditions play an important role."²⁷ On the one hand, it is exactly the notion of abortion as a crime that is criticized by female voices who are underrepresented in the legal discourse: The Club of Progressive Women (Klub kobiet postępowych) comments on the draft law.²⁸ On the other hand, the legal institutions realized the need of linking the law to society's needs. In the 1932 Makarewicz Penal Code, abortion remained criminal, but it allowed an exemption from punishment if the pregnancy threatened the life or health of the woman, as well as for women under the age of 14 and in cases of incest or rape. This juristic progress reflected the democratization of the Polish people in interwar Poland. According to Stephen Robertson, the law is often misunderstood as a tool of the sovereign for ruling the people. Instead, he claims that law also represents the "taught, learned, and accepted customs of people".²⁹ The representatives of Polish Law indeed took their chance to free this field from German, Austrian and Russian influences and adopted it to Polish reality and Polish legal culture.³⁰

---

25 J. Janouli cited from Glaser, Kilka uwag, 1927, p. 37.
26 Grzywo-Dąbrowski, Przerwanie ciąży, 1926 p. 23. Grzywo-Dąbrowski used the term "wives" which explains also his notion of family: a married couple.
27 Czałczyński, Przestępstwo, 1930, p. 4–5.
28 Bujak-Boguska, Pamiętnik, 1930, online: https://polona.pl/item/na-strazy-praw-kobiety-pamietnik-klubu-politycznego-kobiet-postepowych-1919-1930,MTc4NjQzNjU/#info:metadata (05.06.2025), p. 159.
29 Tomlin cited from Robertson, Law, 2005, p. 182.
30 Bołdyrew, Oświata i moralność, 2013, p. 132.

Education Through Advice Literature, or "The Secret Marriage Tricks"

What did advice literature add to the discussions during the interwar period? At the beginning, translated books from Western experts prevailed and circulated widely. In the 19th century, these publications were initially made for a bourgeois readership which can be concluded by the descriptions of homes with several bedrooms which were not the norm for the majority of people. Therefore, it can be assumed that its influence was higher on the upper societal strata as it required not only literacy but also a basic understanding of biology and terminology.[31] When taking a look at the language, one finds astonishing differences: Popular advice literature from abroad can often be characterized as misogynistic. The author of *Secret Marriage Tricks* (*Sekretne sposoby małżeńskie*) claimed that 'wrong intercourse' without reproductive aims was – in moral terms – equal to murder[32] and argued that women requesting contraception were vain because they did not want to harm their beauty by a pregnancy.[33] Popular but ineffective methods like *coitus interruptus* were regarded as the main reason of female hysteria in marriages.[34] Others considered the man to be the sole decision-maker in questions of sexuality: "Regarding sexual intercourse between a man and a woman, it is primarily the man who plays the role of the possessive party, desiring to own the woman physically, while the woman is always the submissive and attractive party to the man."[35]

This attitude was close to the Catholic view which aimed at constructing marriage as the safe and stable core of the country where intercourse exclusively served the purpose of reproduction. Some priests considered contraception to be "national suicide"[36] and an evil promoted by 'the other'. This nationalist conception considered workers returning from France or Germany a threat

---

31   School education was compulsory in Poland from 1921. There are massive differences in literacy between the former partition eras as demographic atlases from that time showed, e.g. Główny Urząd Statystyczny: Rzeczpospolita Polska, 1930. Table 41. https://rcin.org.pl/dlibra/doccontent?id=2495 (05.06.2025).
32   Surbled, Małżeńskie, 1907, p. 8. https://polona.pl/preview/5432af34-fc8a-41ce-9c40-0 d48b91a5dc1 (05.06.2025).
33   Surbled, Małżeńskie, 1907, p. 6.
34   Jackowski, Dzieci, 19XX, p. 20. See also: Surbled, Małżeńskie, 1907, p. 14. Babecki, Eugenika, 1930, p. 24. Gelsen, Nakład Księgarni, 1909, p. 24. Szymański, Ograniczanie urodzin, 1930, p. 31.
35   Eksner, Życie seksualne, 1936, p. 3–4.
36   St. Czarnecki, Zmysły na licytacji, 1932, p. 64.

to the Catholic notion of the family because they brought knowledge about condoms to Polish homes. Besides, Jews were perceived as responsible for the demoralization of the people as they were regarded as profiteers of illegal abortions.[37]

Advice literature genuinely written in Polish developed fast in the 1930s, and was clearly emancipated from foreign convictions. However, it operated with strong metaphors, too. Feminist voices even called human reproduction "blind powers of nature that have to be subjugated (...) in order not to poison someone's life".[38] Justyna Budzińska-Tylicka and Henryk Rubinraut considered motherhood a powerful instinct that places women above men, since women would decide the size of the nation.[39] In their opinion, birth control could turn life into a conscious act by parents who can provide security for their offspring.

Despite the divergent convictions and worldviews, narratives in advice literature always referred to motherhood as the female way to care about the state. The examples tell us more about interpretations of the state of the nation than the actual topic of reproduction. Regardless if a brochure was of progressive or conservative origin, they both spread gender stereotypes (hysteric or passive wife vs. domineering husband). The decision to have or not to have children turned into a national affair.

Real Life Experiences: What Testimonies in Court (cannot) Reveal

The emotional charge of the topic is evident in personal testimonies as presented in the court files, too.[40] Some documents seem to mirror interethnic conflicts in court at first sight, but the following case demonstrates how social pressure influenced the accused woman's testimony. It is about a 21-year-old housekeeper who becomes pregnant from an affair with a Lithuanian farm laborer. The case was heard at the district court of Suwalki in 1928–1929, which means before the law codification. The abortion was performed by a Jewish and a Polish midwife. Seemingly the decision was made by the women herself for

---

37 Renz, planowanie rodziny, 1997, p. 119–120.
38 Ślączkowa, Świadome macierzyństwo, without year, p. 15.
39 Budzińska-Tylicka, Przedmowa, 1933, p. 2.
40 Trials concerning abortion in Poland have been presented in a few case studies: Łodyga, Metody, 2016. as well as Klich-Kluczewska, Przypadek Marii, 2012, p. 195–209.

socio-economic reasons. Although the couple initially presented themselves as engaged, this image was damaged during the interrogation. Marja revised her statement that she wanted the abortion when her partner challenged his paternity in court. "He did not deny that he had sexual relations with (her) but he did not confirm that he was the father of the child either, as other boys had also had relations with her. After the defendant [...] found out that [...] had slandered her, she recanted her original testimony before the prosecutor and gave a second one, indicating that he had persuaded her to induce a miscarriage."[41] Antoni's testimony put Marja in a bad light suggesting a frivolous sexual life for which he would not take responsibility. In consequence, Maria claimed to have been forced to terminate the pregnancy by Antoni. The obligation to live an exemplary life in accordance with societal expectations placed more pressure on women than on men. In some cases, this provoked and chain of accusations and counter-accusations. Both the couple and the midwives were given minor suspended sentences and fines. The midwives appealed unsuccessfully. The behavior of the 21-year-old woman in her testimony to the midwives before the appeal court is worth a closer examination.

"This midwife (points to the first midwife) is a Jewish woman, in whose flat on Jatkowa street the operation was performed. The other one (points to the second) is a Polish-Catholic woman, who, it seemed to me, operated on my genitals. I went to the Jewish midwife (...) on the advice of unknown Jewish women, who I asked on the street in Suwałki"[42] Even if this statement did not have any influence on her own sentence, she seemed aware of the power of her words. First, the deliberate omission of names shows a clear devaluation of the midwives. Second, her statements allude to the notion of a dubious, secretive Jewish abortion network. However, one needs to consider the bigger societal picture before interpreting this statement. In her study about female delinquency in the interwar period, Emilia Płońska states that besides neonaticide, terminations of pregnancy belonged to the most frequently committed crimes of women.[43] Trials of dramatic character were popular among the local population. In the interwar period, there was no selection of participants

---

41   Archiwum Państwowe w Suwałkach (APS): Sąd Okręgowy w Grodnie/Wydział Zamiejscowy w Suwałkach, Serie: 4 Wydział Karny, signature 208, folder 8717, p. 101–102.
42   APS: Sąd Okręgowy w Grodnie. Wydział Zamiejscowy w Suwałkach, Wydział karny, f. 8717, 399. During my research stay in the archive, I could examine 45 cases that were already processed and made available for users in 2019.
43   Płońska, Głośne procesy, 2019, p. 149.

in court rooms, hence the audience consisted of curious neighbors anticipating[44] an unfolding personal drama. The anti-Jewish statement thus could have been done on purpose to distract from her own responsibility and to raise her credibility alluding to the negative image of the Jew the Polish Church was also spreading at that time.

The case cited here was discussed at the county court of Suwałki, therefore one also has to consider the special situation in small towns where everybody knew everybody knew everybody, at least from hearsay as rumors spread quickly. Reports on court trials and unwanted pregnancies and 'fallen women' also were common topics in the local newspapers. The destinies of children born out of wedlock and especially children of prostitutes were a target for catchy headlines (e.g., "Children in the quagmire of fornication").[45] Although analysis of local newspapers also evinced a rising awareness of female living conditions and single mothers, one has to assume the high social pressure and the fear of becoming exposed to the local moral verdict that an accused woman had to withstand.

Conclusion

This article dissected the meaning of unwanted pregnancies on three levels: the legal, educational, and personal. For public discourse, it is worth emphasizing that the opinion of Polish law experts tended to deconstruct the notion of motherhood as a national duty, unlike their foreign colleagues. In Polish advice literature – which certainly had a broader audience than the experts' discourse – the ideological spectrum ranged from normative to progressive; but they were without exception emotionalizing. The connection between motherhood and the state remained crucial. The positions in foreign literature that initially influenced the discussion rather pathologized sexuality without reproduction. Polish opinions, in turn (except the Catholic viewpoint), showed solidarity with female readers recommending they take contraception into their own hands.[46] Considering also the liberalization of abortion regulations unique to Europe at

---

44 Płońska, Głośne procesy, 2019, p. 142. Nonetheless, during my research I came across many court files where a non-public hearing was requested.
45 Dzien dobry Ziemi Suwalskiej, January 8, 1933, p. 9. On the importance of Polish local press See: Piwowarczyk, Model, 2013, p. 131–136.
46 Janicki, Hipokryzji, 2015, p. 340.

that time, I argue that the Polish experts' discourse did not only 'look up' to the Western discourses but emancipated itself from them. Yet the entire codification process is another example of male decision-makers and the way they perceived and judged female topics against the backdrop of the nation building process. For all sources, it is therefore crucial to analyze the narrative potential of the source itself to map the polyphonic reality. Although moral convictions were liberalizing, the language still mirrored a patronizing understanding of women's issues due to the patriotic charge of motherhood.

On the level of individual actors, women were exposed to public opinion that did not change quickly. I argue that the social prejudice about abortions and the unfavorable image of the 'fallen woman' are important to contextualize the spoken word. As the court file showed, negative statements about the Jewish midwife cannot be traced back to actual anti-Jewish beliefs. Primarily, discrediting others during emotionally challenging trials was the last resort to avoid severe judgement both in legal and social terms.

Research with court files thus requires a particular awareness of this type of sources. It is true that law suits on abortion allow us to go beyond the crime itself and to draw conclusions about the personal world. Court files also make voices heard in a way that enables us to visualize former times via listening to normal people "talking about sexual intimacy, power, betrayal, and broken promises."[47] Despite the captivating character of these testimonies, it is important to remember that the sudden character of the statements originates in the heat of the moment. However, they can provide us with insights into the intersections of interests in court and the moral standards according to which the defendants were judged – legally and socially.

Bibliography

BABECKI, Jerzy, Zagadnienie zapobiegania ciąży i eugenika. Odczyt wygłoszony w Sekcji prawno-społecznej Warsz. Oddziału T-wa Eugenicznego, Warszawa, skł. gł. Polskie Towarzystwo Eugeniczne, Bibljoteka Eugeniczna Polskiego Towarzystwa Eugenicznego. Cykl "Eugeniczny", 1930.

BELÁŇOVÁ, Andrea, Anti-abortion activism in the Czech Republic and Slovakia. "Nationalizing" the strategies, in: Journal of contemporary religion, 35 (2020), p. 395–413.

---

47   Robertson, Law, 2005, p. 161.

BOŁDYREW, Aneta, Oświata i moralność. Prawnicy o roli oświaty w procesie modernizacji społeczeństwa polskiego na łamach czasopism prawniczych i spoleczno-kulturalnych na przełowmie XIX i XX w., in: Addenda do dziejów oświaty. Z badań nad prasą XIX i początków XX wieku, ed. Iwonna MICHALSKA/Grzegorz MICHALSKI, Łódź, Wydawnictwo Uniwersytetu Łódzkiego, 2013, p. 131–147.

BUJAK-BOGUSKA, Sylwia (ed.), Na straży praw kobiety. Pamiętnik Klubu Politycznego Kobiet Postępowych, 1919–1930, Warszawa, 1930.

CZAŁCZYŃSKI, Karol, Przestępstwo spędzenia płodu w Komisji Kodyfikacyjnej Rzeczypospolitej Polskiej, Warszawa, 1930.

CZARNECKI, Józef Stefan, Zmysły na licytacji. Szkice pod ostrym kątem, Józef St. Czarnecki., Warszawa, 1932.

DEUTSCHES POLEN-INSTITUT, Jahrbuch Polen 29 (2018). Mythen, Wiesbaden, Harrassowitz Verlag, 2018.

EXNER, Max Joseph, Życie seksualne i miłosne mężczyzny. o czem każdy mężczyzna wiedzieć powinien / M. Dż. Eksner., Warszawa, Wiedza Współczesna, 1936.

GAWIN, Magdalena, The Social Politics and Experience of Sex Education in Early Twentieth-Century Poland (1905–39), in: Shaping sexual knowledge. London, Routledge, 2009, p. 217–235.

GELSEN, Carl von, Hygiena miodowych miesięcy. Wskazówki dla nowożeńców, Gelsen, Wyd. 3 popr., Warszawa, Księg. Popularna, 1909.

GLASER, Stefan & JAKOWICKI, Władysław, Kilka uwag o spędzeniu płodu. ze stanowiska prawa karnego, Warszawa, skład główny w Księgarni Gebethnera i Wolffa, 1927.

GRZYWO-DĄBROWSKI, Wiktor, Przerwanie ciąży z punktu widzenia społecznego, prawnego i lekarskiego, Warszawa, Lwów, Książnica-Atlas, Wykłady Lekarskie, 1926.

HIEMER, Elisa-Maria, Maternity in Times of Crisis. Voices From Interwar Poland on Sexuality and Abortion, Wien/Budapest, CEU Press, forthcoming.

JANICKI, Kamil, Epoka hipokryzji seks i erotyka w przedwojennej Polsce, Kraków, Społeczny Instytut Wydawniczy Znak, Ciekawostki Historyczne, 2015.

KAŁWA, Dobrochna, Kobieta aktywna w Polsce międzywojennej. dylematy środowisk kobiecych, Kraków, Tow. Historyczne Historia Iagellonica, 2002.

Klich-Kluczewska, Barbara, Przypadek Marii spod Bochni. Próba analizy mikrohistorycznej procesu o aborcję z 1949 roku, in: Rocznik Antropologii Historii, 1 (2012), p. 195–209.

Kościańska, Agnieszka, To see a moose. The history of Polish sex education, English-Language edition, New York, Oxford, Berghahn Books, European anthropology in translation 9, 2021.

Kościańska, Agnieszka, Gender, pleasure, and violence. the construction of expert knowledge of sexuality in Poland, Bloomington, Indiana, Indiana University Press, New anthropologies of Europe, 2020.

Krajewska, Joanna, Spór "Wiadomości Literackich" i "Robotnika" o dodatek "Życie Świadome", in: Procesy socjalizacji w Drugiej Rzeczypospolitej 1914–1939. Zbiór studiów, ed. Anna Landau-Czajka, Warszawa, Instytut Historii PAN, Metamorfozy Społeczne 7, 2013.

Kuźma-Markowska, Sylwia, "From 'Drop of Milk' to Schools for Mothers. Infant Care and Visions of Medical Motherhood in the Early Twentieth Century Polish Part of the Habsburg Empire", in: Medicine within and between the Habsburg and Ottoman empires. 18th-19th centuries, papers presented at the conference, Vienna, November 2008, ed. Teodora Daniela Sechel, Gülhan Balsoy, Bochum, Winkler, Das achtzehnte Jahrhundert und Österreich. Internationale Beihefte 2, 2011, p. 131–147.

Kuźma-Markowska, Sylwia, Stan badań nad historią antykoncepcji w XIX i XX wieku, in: Przegląd Historyczny, 100 (2009), p. 603–619.

Landau-Czajka, Anna (ed.), Procesy socjalizacji w Drugiej Rzeczypospolitej 1914 – 1939. zbiór studiów, Warszawa, Instytut Historii PAN, Metamorfozy Społeczne 7, 2013.

Małek, Agnieszka, Z zagadnień historii pracy socjalnej w Polsce i w świecie, Kraków, Uniwersytetu Jagiellońskiego, 2006.

Mędrzecki, Włodzimierz (ed.), Społeczeństwo międzywojenne. nowe spojrzenie, 1 Wydanie., Warszawa, Instytut Historii PAN, Metamorfozy Społeczne 10, 2015.

Michalska, Iwonna/Michalski, Grzegorz (ed.), Addenda do dziejów oświaty. Z badań nad prasą XIX i początków XX wieku, Łódź, Uniwersytetu Łódzkiego, 2013.

Renz, Regina, Kobiety a planowanie rodziny w latach międzywojennych w świetle źródeł kościelnych z Kielecczyzny, in: Kobieta i kultura życia codziennego. wiek XIX i XX. zbiór studiów. T. 5 / pod red. Anny Żarnowskiej i Andrzeja Szwarca. ed. Andrzej Szwarc, Anna Żarnowska, Warszawa, DiG, (Kobieta, Bd. 5), 1997, p. 115–122.

ROBERTSON, Stephen, What's Law Got to Do with It? Legal Records and Sexual Histories, in: Journal of the History of Sexuality, 14 (2005), p. 161–185.

RUSSO, Nancy Felipe, The Motherhood Mandate, in: Journal of Social Issues, 32 (1976), p. 143–153.

SAUERTEIG, Lutz (ed.), Shaping sexual knowledge, A cultural history of sex education in twentieth century Europe, London, Routledge, Routledge studies in the social history of medicine; 32, 2009.

SECHEL, Teodora Daniela/BALSOY, Gülhan (eds.), Medicine within and between the Habsburg and Ottoman empires. 18th -19th centuries; papers presented at the conference, Vienna, November 2008, Bochum, Winkler, Das achtzehnte Jahrhundert und Österreich. Internationale Beihefte 2, 2011.

SIERAKOWSKA, Katarzyna, Elementy kobiecego dyskursu o seksualności na łamach międzywojennych periodyków dla kobiet, in: Kobieta i małżeństwo. społeczno-kulturowe aspekty seksualności. Wiek XIX i XX. zbiór studiów, ed. Anna ŻARNOWSKA/Andrzej SZWARC, Warszawa, "DiG", 2004, p. 365–380.

SIOMA, Radosław/CZARNECKA, Barbara (eds.), Kobiece dwudziestolecie 1918–1939, Toruń, Naukowe Uniwersytetu Mikołaja Kopernika, 2018.

ŚLĄCZKOWA, Zofia, Co to jest "świadome macierzyństwo"? Pewne i nieszkodliwe środki zapobiegające ciąży, Zofja Ślączkowa, Kraków, Magazyn Medyczny Drobner i Sp., 1936.

SLANY, Krystyna, Regulacja prawna aborcji w Polsce w okresie międzywojennym jako element polityki względem rodziny, in: Z zagadnień historii pracy socjalnej w Polsce i w świecie, ed. Agnieszka MAŁEK, Kraków, Uniwersytetu Jagiellońskiego, 2006, p. 139–144.

STAŚKIEWICZ, Joanna, Mutter Polin, Über den Umgang mit einem Mythos, in: Mythen. Jahrbuch Polen 29 (2018), p. 109–117.

SURBLED, Georges, Sekretne sposoby małżeńskie, Warszawa, nakł. Księgarni Popularnej, 1907.

SZLAGOWSKA, Aleksandra, Dyskusje nad świadomym macierzyństwem i dopuszczalnością usuwania ciąży na łamach polskich czasopism medycznych oraz społeczno-kulturowych w okresie międzywojennym, in: Kobiece dwudziestolecie 1918–1939, ed. Radosław SIOMA, Barbara CZARNECKA, Toruń, Naukowe Uniwersytetu Mikołaja Kopernika, 2018.

SZWARC, Andrzej/ŻARNOWSKA, Anna (eds.), Kobieta i kultura życia codziennego. Wiek XIX i XX. Zbiór studiów. T. 5 / pod red. Anny Żarnowskiej i Andrzeja Szwarca., Warszawa, DiG, 1997.

Szymański, Antoni, Ograniczanie urodzin i karalność przerywania ciąży, Antoni Szymański., Lublin, s.n., Bibljoteka Prądu. Serja Nowa, z. 6, 1930.

Walczewska, Sławomira, Damy, rycerze i feministki. kobiecy dyskurs emancypacyjny w Polsce, Kraków, eFKa, Kobieta, Kultura, Krytyka, 1999.

Żarnowska, Anna/Szwarc, Andrzej (eds.), Kobieta i małżeństwo. społecznokulturowe aspekty seksualności. Wiek XIX i XX. Zbiór studiów, Warszawa, "DiG", 2004.

Żarnowska, Anna/Szwarc, Andrzej (eds.), Równe prawa i nierówne szanse. kobiety w Polsce międzywojennej. zbiór studiów, Warszawa, DiG, 2000.

Żarnowski, Janusz (ed.), Państwo i społeczeństwo Drugiej Rzeczypospolitej. zbiór studiów, Warszawa, Instytut Historii PAN, Metamorfozy Społeczne 8, 2014.

Żeleński, Tadeusz, Piekło kobiet, Warszawa, Bibljoteka Boy'a, Bibljoteka Boya, 1930.

Ziegler, Mary, The End of Roe v. Wade, in: The American Journal of Bioethics, 22 (2022), p. 16–21.

Single Mothers and the Issue of Motherhood in Essays and Popular Cinema in Poland in the 1930s

Małgorzata Radkiewicz[1]

Abstract *Małgorzata Radkiewicz analyzes the situation of single mothers in 1930s Poland, drawing on journalistic writings and popular films from that period. In 1918, women in Poland were granted the right to vote, which marked the beginning of a broader public debate concerning the role of women in the public sphere and social institutions, as well as regulations related to medical care and access to abortion. Female authors in particular engaged with issues surrounding gender roles and maternal responsibilities. Since many of these women also wrote screenplays, such themes found their way into genre films—productions that, while conforming to conventional cinematic forms, nonetheless reflected contemporary social realities.*

In 1918 Poland regained its independence, and one of the most important regulations underpinning the new state was the granting of suffrage to all adult citizens regardless of sex. Women gained the right to vote, which they had fought for generations. Women's status as citizens was confirmed by a provision in the Constitution of March 1921, guaranteeing women equal rights in access to public office. In 1929 a new marriage law was drafted, introducing the principle of equality between husband and wife in personal relations. Public debates and the draft new law also included issues of family planning and reproduction, as well as a restrictive provision concerning the criminality of abortion. The restrictive legislation did not take into account the social or family situation of women, which drew sharp criticism from emancipation and liberal circles, and led to a relaxation of the provision in the 1930s: the penalty for a woman deciding to have an abortion was reduced from five to three years, and for the person performing the procedure from a maximum of fifteen to five years. At

---

1 Jagiellonian University, Cracow.

the same time, the code included a provision that abortions could be performed for medical or social reasons. Discussions of family planning and motherhood also addressed the issue of medical care for women and sex education, the need for which was advocated in feature articles at the time.

The changes in the legal status of women affected their social and political activities, which was reflected in their educational and popularization initiatives, artistic work and journalism. As Maria Morozowicz-Szczepkowska, author of theatre plays and screenwriter wrote: "The interwar period was marked by an eruption of female talent in every creative discipline. [...] the already acclaimed Nałkowska, a cool intellectual, other women novelists shone: the talented Kossak-Szczucka, the deeply lyrical Pola Gojawiczyńska, sensitive as a seismograph. [...] there is Magdalena Samozwaniec, with her brilliant caricatural wit in *Na ustach grzechu* (On the Lips of Sin), the good columnist Wanda Melcer and a group of female writers with communist leanings: Krahelska, Szemplińska, Wasilewska. [...] And how many talented feature writers emerged then!"[2]

Morozowicz-Szczepkowska also recalled the community of women centred around the weekly *Kultura Współczesna*, founded in 1927, whose editor was first Wanda Pełczyńska, and from 1934 Emilia Grocholska. As for the magazine itself: "It was a social and literary weekly, dominated by serious, progressive journalism, addressing many current cultural and social issues, persistently fighting for the equality of women in community, social and professional life and the protection of their work. ...Its contributors were Maria Dąbrowska, Zofia Nałkowska, Ewa Szelburg-Zarembina, Pola Gojawiczyńska, Helena Boguszewska, Maria Kuncewiczowa, Hermina Naglerowa, Irena Krzywicka and others."[3]

Interestingly, journalism was followed by other activities: "In 1930, on the initiative of the editorial staff of *Kobieta Współczesna* and the Women's Association, a Club was established in a house at 22 Żurawia Street, on the premises offered for this purpose by Dr. Bronisława Dłuska, the wife of a physician-social activist and a sister of Maria Skłodowska-Curie. It held discussion evenings on scientific, social, cultural and artistic topics."[4]

All these examples show that women's discussions on current social issues and engaged journalism addressed contemporary realities, depicted by the

---

2   Morozowicz-Szczepkowska, Z Lotu ptaka, 1968, p. 269.
3   Ibid.
4   Ibid.

literature of the interwar period. It was believed that films should also deal with everyday problems, providing entertainment and education. The role and functions of Polish cinema were determined by the fact that from 1919 it was controlled by the Ministry of Internal Affairs, which was responsible for the cultural policy of the independent state. It is no wonder, then, that Marian Stępowski, in an article in *Film Polski*[5] in 1923, considered it unnecessary to justify the propaganda role of cinema, concluding that it is a "medium as powerful as the press".[6] As he remarked, after a period of showing the postwar realities and the first years of the reborn state, it was time for a different kind of propaganda: "We must let the whole world know that Poland as a 'young' sovereign state has immeasurable treasures and vital forces ...".[7] This observation can be related to the need to set films in the present with its mores and social circumstances. The text of Stępowski opened a discussion on Polish cinema that would be continued in 1920s and 1930s.

Aniela Waldenbergowa addressed the contemporary women's issue, demanding in *Kino* that such heroines be portrayed on screen: "Everyman – our brother. ... We talk about him, he has passed into literature. He has become a symbol of the crisis, he will go down in history. Somehow no one ever writes about her, about 'everywoman', our sister .... Everywoman takes the burden of maintaining the entire house on her frail shoulders. She works in an office or in a shop. In addition, she looks after the children, often without any help from her partner, who is 'in a rush to get to the office."[8] The lack of time for herself is coupled with a lack of funds, so "everywoman's every dress, item of clothing and footwear is nothing short of a heroic feat! The result of long hours of thinking about where to get money for the material, what to alter, where to find a cheap seamstress?... And she's not the only one who needs clothes, the husband needs to be decently dressed, the children are growing so fast!"[9] Even though the film critic saw *everywomen* as the heroines of their times, cinema was not always interested in portraying women's everyday life.

Film critics publishing in *Wiadomości Literackie* in the 1930s wrote about the necessity of reforming Polish cinema so that it would be closer to reality and ad-

---

5   Stępowski, Pierwiastki propagandy [1923] 2012, p. 39–44.
6   Ibid. p. 39.
7   Ibid.
8   Waldenbergowa, Szara kobieta, 1935, p. 12.
9   Ibid.

dress current problems.¹⁰ The same magazine published at that time texts by Irena Krzywicka and Tadeusz Boy-Żeleński that were notably sensitive to issues of sex education and the need for medical care for women, as well as the consequences of radicalizing abortion law. The presence of contributions from the writer-activists Boy-Żeleński and Krzywicka in *Wiadomości Literackie* was primarily due to the left-leaning character of this literary and cultural magazine whose authors wrote passionately about current issues. For Boy-Żeleński, activism was a writer's duty because he believed that 'literature is a nation's parliament ... It highlights the needs and developments of the hour; it gives voice to the people's demands and wrongs.'¹¹ Commenting on the debate sparked by his writings, he argued in a modernist spirit that the opinions of activists should not be considered only in the context of current problems, but also seen as an indication of future changes in the constantly evolving social order. Researchers of Boy-Żeleński's work emphasize that the author should be considered a moralist, or someone who makes a conscious decision to write about legal and social matters.¹² He chose non-fiction as his weapon for combat in the public sphere, often using maxims, aphorisms, witty comments, and apt caricatures.

Analysis of texts published in *Wiadomości Literackie* at the beginning of the 1930s shows intersections of different kinds between social life and popular culture then in Poland. A similar situation is presented by Shelley Stamp¹³ in her essay on regulating American early birth-control films from 1916 and 1917. She argues that these films "did much more than simply capitalize on a topical, even sensational, issue; they asserted cinema's claim to participate in national debates on an equal footing with newspapers, magazines, and other forms of political commentary".¹⁴ In the conclusion of her analysis, Stamp notes that by the early 1920s, the idea that "cinema's interventionist mandate would be replaced overwhelmingly by visions of its function solely as entertainment".¹⁵ Nevertheless, one might argue that early cinema's role as "an informed public forum"¹⁶ might be followed by individual filmmakers in different historical

---

10   Zahorska, Film polski, 1934, p. 7.
11   Boy-Żeleński, Dziewice konsystorskie, 1958, p. 34. On the role of Boy-Żeleński See also the contributions of Marcin Wilk and Heidi Hein-Kircher in this volume.
12   See Zimand, Boy-moralista, 1958, p. 7–8.
13   See: Stamp, Precaution, 2002, p. 270–297.
14   Stamp, Precaution, 2002, p. 270.
15   Stamp, Precaution, 2002, p. 292.
16   Ibid.

circumstances. In Poland of the 1920s and 30s, as in the United States at the beginning of 20th century, debates over national cinema and the film production system echoed larger questions in the public sphere. Screenwriters and filmmakers tried to incorporate issues of motherhood in popular genre films, particularly melodramas that served entertainment first and education second. This article thus aims at examining how films (and their reception by reviewers) reflected current public debates on women's social and political situation that took place in literature, press and public sphere and will especially take into account how publicists of that time were involved in discussing the gender roles.

Women's Hell

In January 1930, Boy-Żeleński's press columns from the period of October to December 1929 were published as a book about conscious motherhood, provocatively titled Piekło kobiet (*Women's Hell*).[17] In January 1932 *Wiadomości Literackie* carried a review of Boy-Żeleński's book by Paweł Hulka-Laskowski who emphasized that it is 'small in size but immense in content'[18] because it exposes the problems faced by contemporary women and the whole of society. The next issue of *Wiadomości Literackie* featured a note entitled *Głos lekarzy* (The Voice of Doctors)[19] which quoted a letter written to Boy-Żeleński by Zakopane physicians who supported his tireless activity. They declared that his journalistic campaign for the 'improvement of the most neglected areas of life, his uncompromising fight against hypocrisy and reactionism, bold depiction of 'woman's hell', and ... promotion of "conscious motherhood" inspire awe and admiration in sound and independent minds.'[20] The presence of forty signatures under the declaration attests to the deep need for reform felt by the medical community, then unanimous in hoping that the implementation of the ideas promoted by Boy-Żeleński would improve the life and health of all social classes.

---

17   The following bibliographic note appears in the review: Boy-Żeleński, Jak skończyć z piekłem kobiet?, 1932, p. 47. See Hulka-Laskowski, W walce o reformę seksualną, 1932, p. 3.
18   Hulka-Laskowski, W walce o reformę seksualną, 1932, p. 3
19   Głos lekarzy, Wiadomości Literackie no. 3, 1932, p. 5.
20   Ibid.

This debate was waged among journalists as well as physicians and people involved in health care. In the preface to the first edition (there were two more by 1933) of his reflections on 'women's hell,' Boy-Żeleński noted that some of the issues he had raised no longer applied, or that the debate was at a different stage at that moment. However, he emphasized the importance of continuing the debates and literary interventions regarding sexual education: 'Let these notes spread around the world.'[21] He saw the book as an opportunity to maintain the discussion and to remove 'certain intellectual and moral junk'[22] that affected everyday life. In 1933, he stated in the preface to the book's third edition that it had not been possible to change the legislation to mitigate women's hell, though activists and their publications had popularized the idea of conscious motherhood, which in turn had led to the opening of the first family-planning clinic, and, moreover, 'the widest opportunities are available in this respect.'[23] The author also adds that the facts, arguments, and testimonies he had used years earlier remain valid, which confirms the acuity of his judgment and the aptness of his language.

The punishment for abortion, both for women and those who helped them, is the most crucial issue for Boy-Żeleński. The author regards the law as proof of the incompatibility of the legal system with the difficult realities of life where mothers are not provided with appropriate care, assistance, and support. He reflects: 'If you add the cases of young female deaths, cases of severe and permanent disability resulting from the current heartlessly supported state of affairs, perhaps those who draft these laws would shiver in their comfortable seats. And if we included ... suicides, infanticides, and other disasters, we would understand how right it is to call this article "the greatest crime of criminal law."[24] The author writes explicitly about the social hypocrisy that causes women to be left on their own: 'To push a poor girl into motherhood, take her job away because of pregnancy, kick her in disdain, throw on her the entire burden of guilt and threaten her with years in prison [...] – this is the philosophy of laws obviously written by men!'[25]

---

21 Boy-Żeleński, Piekło kobiet. Przedmowa, [1930], 1958, p. 83.
22 Ibid.
23 Boy-Żeleński, Piekło kobiet. Przedmowa, [1930] 1958, p. 85. On the debate regarding 'conscious motherhood' See the contribution of Heidi Hein-Kircher and Marcin Wilk in this volume.
24 Boy-Żeleński, Piekło kobiet. Przedmowa, [1930] 1958, p. 87.
25 Ibid. p. 87–88.

Interestingly, Boy-Żeleński points out the role of men in the situation of women – as legislators, but also as perpetrators of unwanted pregnancies. He does not doubt that, in the absence of care for the mother and child, there is often no way out for the former, especially as the civil code has not regulated the issue of alimony or establishing paternity. So he asks a rhetorical question: 'How can one prosecute only the mother, who bears all the burdens?'[26] He also mentions children born under forced circumstances and deprived of life chances. With the *matter-of-factness* of a physician and social activist he concludes that infanticide rates, especially among unemployed women and low-paid female workers, are depressingly high. The situation of children who will escape that fate is also difficult: 'Such a child will go to the "factory of angels"... where they will likely die. And the ones who have not perished of been raised in the gutter, will join the ranks of the scum of society and increase crime rates. This goes for children born out of wedlock, without a father.'[27] For the sake of clarity, he adds that it is not any easier for legitimate children, as indicated by the horrible situation in workers' flats, filled with numerous offspring living in hunger and squalor.

The July 1932 issue of *Wiadomości Literackie* included 'Życie świadome' (Conscious Living), a special supplement in which Boy-Żeleński[28] explained his metaphor of women's hell. He noted that the hell was still open, especially with the entry into force of the Penal Code on September 1, 1932, which, despite protests, maintained the criminality of abortion. The lost fight of 'reason and humanity'[29] only served to strengthen the attitude of the journalist and doctor, who recalled the history of the Codification Commission's efforts to change the legislation.[30] The new code (adopted in 1932) provided for the punishment

---

26   Ibid. p. 95.
27   Ibid. p. 98.
28   Boy-Żeleński, Piekło kobiet wciąż otwarte, 1932, p. 7.
29   Ibid.
30   Boy-Żeleński gives more details: At the first reading in 1929, the code stipulated a prison sentence of five years for the woman and fifteen years for the person who performed the abortion. As a result of resistance to this provision, an amendment was proposed at the second reading in 1929 that no penalty would be imposed on the physician who performs the abortion if it is necessary due to the mother's health, welfare of the family, or an important public interest'. A clause adopted during the third reading decriminalized the procedure if it was performed on grounds of the pregnant woman's health, her family situation, or public interest as well as the woman's material circumstances. See: Ibid.

of the woman (up to 3 years' imprisonment) and the person who performed the abortion or assisted in its performance (up to 5 years' imprisonment). Abortion was permitted only if the woman's life was threatened or if the pregnancy resulted from one of the crimes listed in the relevant sections. On the one hand, Boy-Żeleński appreciated the fact that at least some exceptions were allowed; on the other, he called attention to the complete omission of the financial aspects and class determinants of women's life. He reiterated that the demands of the Codification Commission included the need to take into account the issues that fell under the concept of family welfare or, more broadly, public interest. The problems of unemployed women who lived in poverty and in large families remained unresolved.

The same supplement to the *Wiadomości Literackie* magazine included an ironic and insightful article by Irena Krzywicka, who pointed out the discrepancy between the modern way of living and thinking and morals – and legislation made mostly by men and the resultant situation of women.[31] According to Krzywicka, the situation is paradoxical because, on the one hand, women have been liberated: 'Woman has relieved man of many obligations. She makes her own living and relies more on her own earnings than on the most tender lover.[32] On the other hand, the social changes have not improved the relations between the sexes or the 'daily interaction of men and women.' She makes an ironic assessment of the situation: 'The current level of seduction is at no more than one metre above the floor, so it does not reach above the waist".

Krzywicka also writes about the brutal and primitive forms of courtship used by men, who 'enjoy only the outcome of love', having no time for a walk or conversation. However, women are not comfortable with this kind of love because it disregards their physical and psychological needs. She describes self-aware women who do not conform to conservative standards and want love, but are condemned to 'austere virtue and sometimes total abstinence' in what she calls a 'society of barbarians'. She adds for the sake of completeness: 'I know women who are prevented from enjoying their erotic life to the full only by their proud aversion to joining the embarrassing race for a disgustingly sated man. There are others, eager and willing, who, in the darkness of the taxicab, are frozen with sadness by the inevitably creeping hand, the only expression of male feelings nowadays.[33] She sums up her bold observations by conclud-

---

31   Irena Krzywicka, Śmierć lowelasa, 1932, p. 7.
32   Ibid.
33   Ibid.

ing that these problems in love life arise from the fact that 'in this area, as in many others, we are in a period of transition which has destroyed the old seducers, lady-killers, and Don Juans, but has failed to produce their replacements, which has not given rise to a new culture or a new love ritual.[34] The future may bring changes, but only if women themselves are involved in them. So she appeals to women: 'But men are not going to do this for us; we women must rely only on ourselves ... We cannot look to the church or bourgeois "principles" for regulators of our conduct: we must find them in ourselves, in our conscious intention, not motivated by calculation or absurd snobbery.[35]

Theatre and Film (Melo)Dramas

Krzywicka was not the only one who argued against patriarchal social structure. A similar attitude could be found in the writings of Maria Morozowicz-Szczepkowska, an author of screenplays and theatre dramas, who, as Boy-Żeleński and others, openly addressed the issue of motherhood and being a single mother. In 1933 her play entitled *Milcząca siła* [The Silent Power] staged at the Great Theatre (Teatr Wielki) in Warsaw was reviewed by Henryka Felkowska as "the latest female art".[36] According to Felkowska, the viewers were proposed "three hours of a discussion on stage" about the contemporary woman – her rights and emancipation tendencies, especially that "Morozowicz-Szczepkowska, the author of a famous and controversial drama *Sprawa Moniki* [Monika's Case, 1933] about a love triangle, has gone even further in her feminism. While the first play wanted to prove that man is not indispensable for life [...] *The Silent Power* tries to convince us that in the future, better world, men will simply be redundant".[37]

Moreover, Felkowska argued that if the message of the play may be considered disputable, its concept might cause interest as the vision of the future in which a woman – an editor in an influential women's magazine, managing the team of "women of all races and nationalities" – does something possible to make this "silent power" gain power over the world just for the good of this world. In so doing, she follows her conviction that "Humanity infected by war

---

34   Ibid.
35   Ibid.
36   Felkowska, Mężczyzna, 1933, p. 2.
37   Ibid.

and crime of every description is heading towards destruction and it is only a change of the regime that can save it".[38] In the play, the currently gathering "World Women's Congress" is expected to have a decisive meaning, and the editor was to make a speech there as an activist and single mother. To achieve this goal, she had to struggle against representatives of male power, as well as equally obstructive representatives of "women of the old school". As the writer of the review ironically observes: "What a keen observation: He will always have followers and defenders among the ranks of 'the silent power".[39]

The way Felkowska presented the ideas of Morozowicz-Szczepkowska leaves no doubt that as a writer she was familiar with the ideas of Boy-Żeleński and Krzywicka, and shared with them the same attitude towards the realities of Polish women. As Felkowska remarked in her interpretation of *The Silent Power*: "the playwright's main thesis is [...] very extreme, although not so new. Recently male voices, even among scientists, have appeared which attributed the reason of the fall of humanity to the fact that it is only men who govern it [...] (please see Professor Bergman's book *Erkenntnisgeist u. Müttergeist*).[40] So they sought remedy in the return to matriarchate."[41] According to the reviewer, the best solution would be the harmonious government of both sexes which, however, is difficult to see while watching the play because "the female characters are alive and active. Men, unfortunately, are shaped following a demanded policy".[42] Nevertheless, the reviewer finds the idea itself very interesting, namely the introduction of a feminist theme on theatrical stage. This is especially so because "the discussion of this type on the theatrical stage is a novelty in our country, and thanks to the fact that there are very sincere and very strong moments in it, the play is neither tiring nor tiresome, on the contrary – it intrigues and generates interest".[43]

Judging by the review of her play, it is not surprising that Morozowicz-Szczepkowska was willing to participate in the discussion on "women's hell". Her involvement and critical attitude are reflected in a screenplay for the most interesting film about unintended pregnancy and lack of support for women – namely, *Wyrok życia* (*Life Sentence*), directed by Juliusz Gardan in 1933. The

---

38 Ibid.
39 Ibid.
40 Bergmann, Erkenntnisgeist, 1932.
41 Felkowska, Mężczyzna, 1933, p. 2.
42 Ibid.
43 Ibid.

screenwriter's feminist view and radical opinions about social and moral issues related to women's life were reflected in the film's working title *Kto winien?* (*Who's Guilty?*), which sounds like a quote from Krzywicka's committed essays or Boy-Żeleński's columns. Most of the film's action takes place in a courtroom where a young woman is on trial for killing her newborn child. The only person willing to help her is her lawyer – the sole female member of the bar – who recognizes the difficult situation and living conditions of women, especially when they not only lack the support of their families, but also have no work or income. The defence establishes that the child died accidentally when the mother wanted to commit suicide. The baby fell into the water instead of the mother and could not be saved. The scenes in which the woman is stigmatized, reminiscent of lynching, are juxtaposed with images of her loneliness and rejection when it transpires that the affair has left her pregnant. The man has disappeared from the woman's life and the illegitimate child only reinforces the mechanism of exclusion caused by her economic status.

When the advocate finally manages to secure freedom and an opportunity for rehabilitation through work for the woman, it turns out that the girl's pregnancy and troubles were caused by the lawyer's husband. Upon learning the whole story, he tries to commit suicide; after he is saved, he confesses his love to his wife who decides to stay with him and lets the woman go. This ending matched the film's melodramatic tone, but was far from the intention of the screenwriter who wanted the women to reject the man and remain friends to support each other in later life, as did the heroines of her play *Sprawa Moniki* [Monika's Case].

Life Sentence addresses the themes discussed by Boy-Żeleński and Krzywicka, but also issues related to women's rights and their position in public life. They were noted in reviews of the film, which emphasized its timeliness and social impact. Jim Poker (alias Julian Ginsbert, a writer and a journalist) wrote enthusiastically in the magazine *Kino*: 'So let me celebrate a little, because I was right. It turns out that it is possible to make a good Polish film [...] in which an ingenious script with a noble message, combine with competent performances and the chief asset. [...] the whole is artistic and original, it makes use not of antiquated clichés, but of a beautiful idea, allowing the audience for an hour and a half to live, suffer, be moved, and rejoice with the characters. *Life Sentence* is not only well paced, but, above all, meaningful. Its meaning is not superficial, but deep and social. The meaning does not insinuate itself in the form of a moral, but every spectator senses it perfectly'.[44]

In his analysis Poker emphasized both the artistic and social values of the film: "A wonderful opening ... The court, an excellent jury, screaming women, the interior of a printing works, spectacular and numerous outdoor locations, ... erotic scenes shot discreetly, with great tact and restraint and, at the same time, with realism and simplicity, without trying for theatrical demonism, brilliant supporting characters, a profound and truly dramatic scene of childbirth, intensified by the juxtaposition with nature (wind)".[45]

It is interesting to compare the review with pre-production discussions and comments. In 1933, when Gardan's film was still in production, the magazine *Wiadomości Filmowe* [*Film News*] announced it as an interesting project: 'The subject of the film *Who's Guilty?* [working title later changed to *Life Sentence*] is rooted in life, yet completely new and original. A young girl has killed her child, the fruit of an enchanting night of love that was followed by a rude awakening. The seducer has disappeared and faces no consequences while a series of misfortunes befalls his victim. This serves as the basis for a gripping tragedy of love. The film is a fervent protest against all social injustice.'[46]

After the film's premiere, *Wiadomości Filmowe* wrote once again about *Life Sentence* in *the Talk of the Day* column. It reads: 'If you listen to people talking in cafes, sweet shops, dance halls, trams and taxis, it turns out that the most common topic of conversation now is *Life Sentence*, the latest production of Blok-Muzafilm."[47] The title of a note on the same page of the magazine read: "Four Aces in One Hand. 'It's easy to win a game with cards like these ... Who are the aces? One is Maria Morozowicz-Szczepkowska, the author of the record-breaking *Monika's Case*. ... First Polish screenwriter returned to film after a few years and wrote a sensational script called *Life Sentence* (*Who's Guilty?*). ... The film's main actors are the other three aces. ... Namely: Jadzia Andrzejewska, Irena Eichlerówna and Dobiesław Damięcki. [...] The film's subject – social injustice in matters of love and morality is strikingly original."[48]

The problems addressed by the film are given even more resonance on the same page in a piece entitled 'In Defence of Unwed Mothers', pointing out, as

---

44 Poker, Wyrok życia [review], 1934 p. 8–9.
45 Ibid.
46 Wiadomości Filmowe, 1933, no. 10, p. 2.
47 Wiadomości Filmowe 1934, no. 1, p. 3
48 Ibid.

Boy, the role of men in the situation of women: 'When a young girl succumbs to the seducer's sweet words and is then abandoned with a living memento of her passion – the full weight of moral outrage falls on her. When the child is lost, she becomes a suspect and faces terrible legal consequences. But the real culprit is someone else! Who? The answer to this intriguing question can be found in the latest picture by Blok-Muzafilm, *Life Sentence* [...]. A group of experts who have already seen *Life Sentence* unanimously regard these two roles as the greatest in the history of Polish cinema. The dramatic conflict between the two heroines is caused by the husband of one and the lover of the other. He is played by Dobiesław Damięcki, the first Polish screen lover of a new kind. The film promises to be the season's biggest sensation and has already attracted a lot of attention.'[49]

Moreover, *Wiadomości Filmowe* printed a very original advertisement for *Life Sentence* in a box in which the names of cinemas that showed the film were accompanied by a list of reasons for seeing it:

> '*Life Sentence* is a sentence against male seducers and a defence of seduced girls.
> *Life Sentence* is an eternal problem of man and woman in the context of the present day.
> *Life Sentence* is a film about an erotic tragedy of unprecedented depth ...
> *Life Sentence* is a great protest against injustice to women...
> *Life Sentence* is a great force paving the way for a better tomorrow.'[50]

The list could be easily elaborated in a form of an essay, using arguments of Boy-Żeleński, Krzywicka and others. Unfortunately, Polish popular cinema did not follow the path of social activism (the same as the American film industry described by Stamp), submitting its function to entertainment. The best example of commercialization of the issues of motherhood and being a single mother could be *Serce matki* (A Mother's Heart), a popular novel by Antoni Marczyński, which was also adapted for the screen. The director Michał Waszyński initially intended to call the picture *Macierzyństwo* (*Motherhood*), but the title must have been considered insufficiently catchy and distant from the literary original because the film was released as *A Mother's Heart*.

---

49   Ibid.
50   Ibid.

The magazine *Film* wrote that it is a story of 'two women fighting for the right to the child of the man they both love.'[51] The melodramatic plot tells of a young teacher who becomes pregnant after a short affair. The female doctor who looks after her turns out to be the wife of her lover so the future mother agrees to give the child away to her to provide it with a father and a new guardian as well as social stability. She takes up a job as a cleaner in the kindergarten to be closer to her daughter. The man doesn't know anything and looks after the child as if it were his own and his wife's; when he learns the truth, he wants to be reunited with his beloved but dies in a car crash. The two women decide to join forces and raise the girl together. At least the theme of educated, working women who can meet the challenge of raising a child, stands out against the melodramatic formula. The motif of female friendship is illustrated by the situation of the woman whose flatmates – also employed, independent and single – are willing to help her in her maternal duties. They even buy a pram, but the protagonist lies to them that the child died in hospital.

Yet, the atmosphere of the discussion on motherhood and women's lives from the beginning of 1930s was reflected in another film adaptation based on a novel by Pola Gojawiczyńska, *Dziewczęta z Nowolipek* (The Girls from Nowolipki, 1935), who carefully described the social situation of women in the interwar period. Her complex novel portraying a group of young girls living in the same tenement in a poor district of Warsaw was filmed by Józef Lejtes in 1937. By setting the story in a single building with a closed-in courtyard, both the writer and the director were able to emphasize the issue of social inequalities affecting the girls' lives. Growing up, young women would experience social rejection for economic and class reasons and on mora grounds due to the public perception of single women's affairs. They would also suffer disappointment in love and unwanted pregnancies. The structure of the novel, rich in plot and characterization, was kept in the film by Lejtes, providing diverse portraits of women, sensitively played by female film stars, including Jadwiga Andrzejewska, know from *Sentence of Life*. Both the cast and the excellent portrayal of realities must have contributed to the film's success.

The examples provided of works by Boy-Żeleński and female authors such as Krzywicka, Gojawiczyńska, and Morozowicz-Szczepkowska prove that the beginning of 1930s in Poland was a time of women's liberation, when women who had gained the right to vote more than a decade earlier, tried to obtain

---

51   See [Review] Film 1938, no. 32, p. 21.

better access to systems of health care and sexual education. It was the question of taking women's rights seriously in social life at large, but also in literature and popular cinema. As Krzywicka argued in her writings, modern women should work for her independence, breaking social and cultural taboos (pregnancy, maternity, abortion) and liberating her body from social conventions (of women's sexuality and gender roles in particular). She discussed the idea of a liberal society that equally treats both sexes, encouraging women to free themselves from patriarchal tradition and social patterns. Echoes of these ideas might be found in popular cinema, although transformed and subordinated to film genres.

It is interesting to examine interwar writings from the perspective of contemporary Poland. In *Women's Hell* Tadeusz Boy-Żeleński stressed that the issues of sex education, especially conscious motherhood and birth control, 'are a thing of the future, although not a distant future. We are living at a time when a new penal code is being drafted in Poland. Tormented by law and current ethics, woman cannot console herself that the next generation will cast off these chains. It is therefore necessary to raise a cry for justice for today's women.'[52] History has proven very perverse and shown that the fight still goes on, and the 'future' described in Boy-Żeleński's columns turned out to be the present in the autumn of 2021 when a wave of protests connected with the women's strike swept through Poland. The most common slogan on posters and banners was 'women's hell', which continued to be the reality for Polish women in the early 21st century.

Bibliography

BEAN, Jennifer M./NEGRA, Diane (eds.), A feminist reader in early cinema, Durham, London, Duke University Press, A camera obscura book, 2002.

BERGMANN, Ernst, Erkenntnisgeist und Muttergeist: eine Soziosophie der Geschlechter, Breslau, 1932.

FELKOWSKA, Henryka, Mężczyzna nie jest potrzebny... Trzy godziny scenicznej dyskusji w nowej sztuce kobiecej Morozowicz-Szczepkowskiej, in: Awangarda 8, 1933, p. 2.

GIERSZEWSKA, Barbara (ed.), Polski film fabularny 1918–1939: recenzje, Kraków, Księgarnia Akademicka, 2012.

---

52 Boy-Żeleński, Piekło kobiet. 1932, p. 173.

HULKA-LASKOWSKI, Paweł, W walce o reformę seksualną, Wiadomości Literackie 2 (1932)2, p. 3

KRZYWICKA, Irena, Śmierć lowelasa, in: Wiadomości Literackie, 32 (1932).

MOROZOWICZ-SZCZEPKOWSKA, Maria, Z lotu ptaka: wspomnienia, Warszawa, Państwowy Instytut Wydawniczy, 1968.

POKER, Jim, Wyrok życia [review], in: Kino 1 (1934).

STAMP, Shelley, Taking Precaution, or Regulating Early Birth-Control Films, in: A feminist reader in early cinema, ed. Jennifer M. BEAN/Diane NEGRA, Durham, London, Duke University Press, A camera obscura book, 2002, p. 270–297.

STĘPOWSKI, Marian, Pierwiastki propagandy państwowej w filmach polskich. Z powodu obrazu pt. "D'Elmoro-walka o skarby" [Film Polski 1923, no. 2–3], in: Polski film fabularny 1918–1939: recenzje, ed. Barbara GIERSZEWSKA, Kraków, Księgarnia Akademicka, 2012, p. 39–44.

WALDENBERGOWA, Aniela, Szara kobieta. Bohaterka naszych czasów, in: Kino: tygodnik ilustrowany, 6 (1935), p. 12.

ZAHORSKA, Stefania, Film polski w impasie, in: Wiadomości Literackie 50 (1934), p. 7.

ŻELEŃSKI, Tadeusz (Boy), Dziewice konsystorskie, Warszawa, Księgarnia Robotnicza, 1929.

Żeleński, Tadeusz (Boy), Piekło kobiet, Warszawa, (Bibliotheka Boy'a), 1930.

ŻELEŃSKI, Tadeusz (Boy), Piekło kobiet wciąż otwarte, Wiadomości Literackie 32 (1932).

Żeleński, Tadeusz (Boy), Jak skończyć z piekłem kobiet? („Świadome macierzyństwo"). Warszawa (Biblioteka Boya), 1932.

ŻELEŃSKI, Tadeusz (Boy), Pisma 15: Dziewice konsystorskie. Piekło kobiet. Jak skończyć z piekłem kobiet? Nasi okupanci, ed. Henryk Markiewicz, Warszawa: Państw, 1958.

ZIMAND, Roman, Boy-moralista, in: Pisma; 15: Dziewice konsystorskie. Piekło kobiet. Jak skończyć z piekłem kobiet? Nasi okupanci, Warszawa: Państw. 1958, p. 7–8.

"War of births"
Midwifery under German Occupation in the Wartheland, 1939–1945[1]

Wiebke Lisner[2]

Abstract *This article is dedicated to the question which roles and functions German and Polish-Christian midwives assumed in occupied Poland (1939–1945). After the military conquest of Poland, Nazi health policy-makers saw themselves as waging a 'birth war'. German midwives, as experts in birth and reproduction were given a specific role in this. However, the help of Polish midwives could not be dispensed with. What options for action did this create for midwives and mothers?*

Nazi Germany invaded Poland on September 1st, 1939. A few weeks later, on October 26, the occupied western Polish territories were annexed to the German Reich as Reichsgaue Posen (later Wartheland or Warthegau) and Danzig-West Prussia, and as the government districts of Kattowitz and Zichenau. The occupied central Polish territories were placed under German rule and administration as General Government. In the annexed western Polish territories, some 10 million former Polish citizens came under German rule. The newly formed Reichsgau Wartheland, with its main cities of Poznań/ Posen and Łódź/ Lodz/

---

[1] The text is based on my Habilitation thesis, which was accepted by Hannover Medical School and published 2025: Hebammen im Wartheland. Geburtshilfe zwischen Privatheit und Rassenpolitik, 1939–1945.
[2] Institute for Ethics, History and Philosophy of Medicine, Hannover Medical School.

Litzmannstadt³ contained about 4.5 million people, of whom 88% were Polish Christians, 9% Polish-Jewish and 7% "ethnic Germans".⁴

The aim of the Nazi occupation regime was to "Germanise" the annexed western Polish territories by reshaping the population and transforming the infrastructure and public life.⁵ This involved the settlement of ethnic Germans from south-eastern and eastern Europe, the exclusion and deportation of Poles and Jews, and a population policy that encouraged the birth and care of German children and discouraged the birth of "foreigners".⁶ Moreover, after the military victory over Poland, Nazi health policymakers saw themselves as engaged in a "war of births" with the Poles. They singled out Polish women in particular as opponents who sought to win the "war" biologically and thus countering the German aim of Germanising the territory.⁷ Nazi health policymakers regarded enabling (pronatalist) and disabling (antinatalist) population policies as crucial to winning the biological "Volkstumskampf" (ethnic struggle) in favour of the Germans.⁸ Thus, reproduction and reproductive practices became a major target of occupation and Germanisation policies in the annexed western Polish territories, based on the ideas of eugenics and race hygiene, with its evaluative logic at its core and its specific Nazi racist orientation.⁹ Arthur Greiser, the Reich Governor of the Warthegau was a radical proponent of the policy of "Germanisation". He wanted to turn the Warthegau

---

3   The German authorities changed all Polish names into German for the aim of Germanising the territory. Cf. Haar, Genozid, 2009. In the following, when writing about western Poland under German occupation, I use the German names.
4   Cf. Pohl, Reichsgaue, 2007. Stiller, Politik, 2022, p. 348.
5   Haar, Genozid, 2009.
6   Cf. Stiller, Politik, 2022.
7   Grohmann: Erb- und Rassenpflege als Grundlagen biologischer Volkstumspolitik (Hereditary and racial tending as the basis of biological ethnic politics), 7 Oct. 1941. In: Archiwum Państwowe w Poznaniu (APP), 299 RSH/1137, p. 33.
8   Cf. Vossen, Typen, 2013.
9   By the end of the 19th century, eugenic ideas and the goal of "breeding" humans had spread throughout the world. Cf. Bashford, Levine, Eugenics, 2010, p. 3–24. Under National Socialism, eugenics was linked to racial policy, which was elevated to a state premise and became the defining political element, developing its specific radicalism under the conditions of dictatorship. Cf. Schmuhl, Entwicklungsdiktatur, 2009. On the connection between population policy, racial policy and a radical Germanisation policy during the Second World War, see Haar, Genozid, 2009.

into a "model Gau" of Germanisation and promoted a radical Germanisation policy.[10]

However, pregnancy and childbirth affected and still affect women and their bodies, involving personal choices and emotions, and thus refer to a more private sphere, but also to social norms and values, and religious beliefs.[11] In the 1930s and 40s in both Poland and Germany, most births took place in women's homes, attended by a midwife.[12] Midwives thus had a central position in the female-dominated reproductive sphere of pregnancy, childbirth and early parenthood. The decentralized organisation of obstetrics allowed midwives and women to act autonomously. At the same time, the importance of reproduction for population policy meant that pregnancy and childbirth were of the utmost interest to the state.[13] The German health authorities assigned German midwives, as experts in pregnancy, childbirth and postnatal care a "war-decisive" key position in the annexed Polish territories. They were expected to educate German mothers in order to improve their health and that of their children.[14] The educational work of the midwives however, was not only intended to reduce maternal and infant mortality, but also to create an "extended ethnic community" (Volksgemeinschaft) in the annexed Polish territories.[15] As a 1939 memorandum from the "Rassenpolitisches Amt" (Office for Racial Policy) in the Warthegau put it, the goal was "to create a racially and thus spiritually and politically united German population" in the annexed western parts of Poland.[16]

Despite radical Nazi propaganda promoting a "war of births" and racial segregation policies, maternity care in the Warthegau was not possible without the help of Polish midwives. In 1941 there were around 800 Polish midwives

---

10   Cf. Epstein, Model Nazi, 2010. Alberti, Exerzierplatz, 2006.
11   Herzog, Reproduction, 2018.
12   In 1939 in the German Reich about 61% of all births took place at home. In occupied Poland, in the district of Posen, in 1943 freelance midwives attended 93% of all births. Cf. Lisner, Hüterinnen, 2006. Midwifery Statistics, 1943, in: APP, RSH 299/1920, p. 19–20.
13   Cf. Lisner, Hüterinnen, 2006. König, Wacht, 2024.
14   Witte, Haupttagung, 1940, p. 228–229.
15   Cf. Kundrus, Regime, 2009.
16   Hecht, Wetzel, Behandlung, 1939, in: National Archives and Records Administration (USA), RG 238, PS-660. I thank Alexa Stiller for letting me use the source.

practising, compared with only 146 German midwives.[17] How did this clash between population policy goals and their practical implementation affect the scope of action of midwives and mothers?

While Nazi Germany's birth policies and midwifery in the "old Reich" (Germany within its 1937 borders) have been well researched,[18] there have been few efforts to explore the means of reproduction and the role and function of midwives in the context of German occupation policy.[19] The following article uses sources from Polish archives and midwives' reports published in the Nazi midwives' journal to analyse the roles and functions of German and Polish midwives in the Nazi "war of births" in the Wartheland. The article thus contributes to the history of reproduction under National Socialism and of midwives in occupied Poland.

Population and Birth Policy in the Warthegau

Hitler proclaimed that the purpose of the invasion of Poland was to create a new "living space" for the Germans.[20] As Hitler explained in his speech to the Reichstag in Berlin on October 6, 1939, he envisioned a "new order of ethnographic relations, that is, a resettlement of nationalities".[21] "Fragments of German nationality" living outside the German Reich without German citizenship were to be "resettled" (rückgesiedelt) into the German Reich.[22] On October 7, Hitler entrusted the implementation of the "Volkstumspolitik" to Heinrich Himmler, who gave himself the title of "Reichskommissar für die Festigung des deutschen Volkstums" (Reich Commissioner for the Strengthening of the German Nation, RKF). Contrary to the use of the term in the early 1930s, the RKF apparatus no longer understood "Volkstumspolitik" to mean only the promotion of "ethnic German" groups outside the German Reich. Rather, "Volkstumspolitik" now included the exclusion of "foreigners" and "undesirable" population groups through expulsion, deportation and murder, in order to create

---

17 Abteilung II, Lagebericht der Abteilung II für das 4. Vierteljahr 1942, 20 January 1943, in: APP, 299/1880, p. 65–66; Berichterstattung über die gesundheitliche Betreuung der Zivilbevölkerung, hier: Hebammeneinsatz, 15 January 1944, in: APP, 299 RSH/2031.
18 Cf. Peters, Biographie, 2018. Lisner, Hüterinnen, 2006.
19 Cf. Haar, Genozid, 2009. Esch, Vehältnisse, 1998. Lisner, Hebammen, 2016.
20 Wildt, Haupttagung, 2007.
21 Hitler quoted after Wildt, Haupttagung, 2007.
22 Ibid.

areas for German settlement and to secure conquered territories.[23] Approximately one million "ethnic Germans" from south-eastern and eastern Europe were resettled to the "incorporated" western Polish territories, especially to the Warthegau, and thus contributed to the "Germanisation". At the same time, about 281.000 Poles and Jews were deported to the General Government, another 254.000 were forcefully relocated within the Wartheland and about 12% of the Polish population was deported to the German Reich for forced labour. However, the Poles remained the largest group in the Wartheland until the war ended.[24]

In addition to the settlement of ethnic Germans and the exclusion and expulsion of Poles and Jews, Greiser implemented measures to increase the German population. In October 1939, he introduced the "Deutsche Volksliste" (German Ethnic Register) to select Poles and Germans – Jews were completely excluded. In the Wartheland, about 12% of the former Polish citizens were included in the German Ethnic Register. Those classified as "German" were assigned to four different categories. The grouping of the German population in the categories I-IV was linked to the granting of varying degrees of citizenship rights.[25] The German Ethnic Register thus flexibilized the boundaries of Germanness.[26]

However, as some health experts, such as Herbert Grohmann, medical officer in Litzmannstadt, pointed out, the "war of births" remained a problem for the successful Germanisation of the Wartheland.[27] In 1941 Grohmann noted that the Poles had lost the battle of arms, but were determined to win the "national struggle" through their greater biological strength, as evidenced by high marriage and birth rates.[28] Grohmann called for the "biological strength" of the Polish people to be broken.[29] In an effort to keep the number of children in the Polish population as low as possible, on September 10, 1941 Greiser introduced

---

23   Stiller, Politik, 2022, p. 15.
24   Cf. Stiller, Politik, 2022, p. 408–411. Linne 2013.
25   The Germans were categorised based on judgements about their past commitment to "Germandom", their ethnic ancestry and their potential social value as Germans. Especially those assigned to the groups III and IV had to demonstrate their Germanness and had to obey special regulations and restrictions. Cf. Heinemann, Rasse, 2003. Wolf, Herrschaftsrationalität, 2012.
26   Kundrus, Regime, 2009.
27   Cf. Vossen, Typen, 2013.
28   Grohmann: Erb- und Rassenpflege, 1941, p. 33.
29   Grohmann: Erb- und Rassenpflege, 1941, p. 33–34.

a minimum marriage age of 28 for Polish men and 25 for women. In addition, abortion was encouraged among Polish women as long as it was approved by a German medical board.[30]

According to Grohmann, in order to win the "national struggle" it was also necessary to achieve a "numerical strengthening of the German ethnic group".[31] A report by Kurt Schmalz,[32] deputy Gauleiter of the Reichsgau Wartheland, underlined this aim and specified the measures to be taken. He called for the prevention of miscarriages among German women and for better care for German mothers and infants to reduce the death rates caused – as he analysed – due to improper treatment and lack of education. As a key element in achieving this goal, he demanded "the elimination of all deficiencies in the field of obstetrics, maternity and infant care" – of course, only for the German population. In addition to the lack of medical care, Schmalz saw the "wrong attitude" and "ignorance" of ethnic German mothers as the main causes of the high mortality rates among German infants. The report emphasized that the mortality rate could be reduced through education in health-promoting behaviour.[33]

Midwives Put Central Stage at the "War of Births"

German midwives played a key role in implementing the Reichgovenor's plans to reduce maternal and infant mortality. Schmalz called for practical training of German mothers by German midwives.[34]

---

30   Cf. Epstein, Model Nazi, 2010, p. 201, 216. Majer, Fremdvölkische, 1981, p. 257, 413.
31   Grohmann: Erb- und Rassenpflege, 1941, p. 28.
32   Kurt Schmalz (1906–1964) was deputy Gauleiter of the Warthegau from March 10, 1941–1945. Previously, he had been deputy Gau Leader in the Gau of South Hanover-Braunschweig. In the Warthegau, Hitler appointed Arthur Greiser from Danzig as Reichsstatthalter and Gauleiter. Cf. https://de.wikipedia.org/wiki/Kurt_Schmalz (05.06.2025). https://www.zukunft-braucht-erinnerung.de/uebersicht-der-nsdap-gaue-der-gauleiter-und-der-stellvertretenden-gauleiter-1933-1945/ (05.06.2025). For Arthur Greiser Cf. Epstein, Model Nazi, 2010.
33   Report of the deputy Gau Leader in the Warthegau, Ne/Wa., 1 Oct. 1941: concerning Midwife outreach in the Wartheland, in: IPN, GK 746/66, p. 89.
34   Report of the deputy Gau Leader in the Warthegau, Ne/Wa., 1 Oct. 1941: concerning Midwife outreach in the Wartheland, in: IPN, GK 746/66, p. 104.

As in the "Old Reich", the German health administration in the Warthegau relied on decentralised midwife-led obstetrics. Low hospital density, a shortage of beds and a lack of transport made any consideration of centrally organised maternity care in clinics null and void. In addition, the Reich Health Leader, Leonardo Conti, insisted that home births be encouraged in the Warthegau, as in the "Old Reich", in accordance with the Decree on the Promotion of Home Births of 1939.[35]

From the autumn of 1939, the Reichshebammenschaft mobilised German midwives from the German Reich for temporary or permanent work "in the East" as part of the resettlement of ethnic Germans and to care for the German population in the settlement areas. One of the German midwives who volunteered was Margarete Riedel, a midwife from Berlin. After training at the midwifery school in Berlin-Neukölln, she settled in the district of Posen. She wrote in the National Socialist midwives' journal:

> "My work here gives me great pleasure [...] My mothers often include resettled women [...] Thanks to the excellent training by Prof. Ottow [...] it is easy for me to give advice and help."[36]

Like many of her colleagues, Margarete Riedel was enthusiastic about working in the "incorporated areas". In their reports, German midwives from the German Reich emphasised the independence of their work, the sense of community among the German population and the flat hierarchies in relation to doctors.[37] Within the racist social hierarchy established by the German occupation administration, with Reich Germans at the top, an intermediate ethnic German class, Poles as a labour reservoir and Jews at the bottom, they enjoyed privileges as the "master race" and experienced the Warthegau as a space of almost unlimited possibilities.[38] Gender and social hierarchies that had existed in the "Old Reich", such as those between doctors and midwives, seemed flatter to them because of the superimposition of the racist social hierarchy; the pronounced sense of community within the German population and the promised

---

35 The decree was partly enacted to keep clinics free for wounded soldiers. Cf. Lisner, Hebammen, 2016.
36 Riedel, Eindrücke, 1941. Prof. Benno Ottow (1884–1975) was head of the midwifery school Berlin-Neukölln and closely connected with Nanna Conti. Hansson, Peters, Tammiksaar, Sterilisierungsoperateur (2011).
37 Cf. Morsbach, Erlebnisse, 1942. Prinz, Erlauschtes, 1943.
38 Cf. Röger, Kriegsbeziehungen, 2015.

"extended Volksgemeinschaft" were within reach.[39] In addition, the health administration offered a minimum wage twice as high as in the "Old Reich" as well as housing and financial assistance to buy equipment and a car.[40]

It was precisely this sense of expanded opportunity and self-determination described by the midwives, that Nazi planners sought to evoke in the German population in order to make settlement "in the East" a success. In their November 1939 memorandum of the NSDAP's Office for Racial Policy (Rassenpolitisches Amt der NSDAP), the two lawyers Erhard Wetzel and Günther Hecht stated: "The Germans in these areas must be given the feeling of more space and personal freedom of development, both in the city and in the country".[41]

However, there were not enough German midwives to provide comprehensive care for the Germans. According to the 1941 report by Schmalz's office, there was a shortage of about 400 German midwives to provide at least one in every district.[42] As a result, German mothers were almost exclusively dependent on Polish midwives or the help of a neighbour.[43] However, even if most midwives were still Polish in 1944, the recruitment of German midwives was somewhat successful: By 1943 the number of German midwives had doubled from 146 in 1941 to 300. Most of the midwives came from the "Old Reich" seeking new opportunities in occupied Poland or because they couldn't get permission to practice in their home districts.[44] In addition, some of the German midwives

---

39 Cf. Morsbach, Erlebnisse, 1942. Morsbach, Arbeit, 1942. Elizabeth Harvey concludes this for many women employed in the "Nazi East". Cf. Harvey, Women, 2003.
40 Cf. Gebührenordnung (fee schedule), 14 June 1941, in: APP, 299 RSH/2070, p. 11–112. In the Wartheland, the occupation authorities often offered the best housing of the former Polish and Jewish inhabitants they had expelled to the Germans from the Reich. Cf. Kehle, Hebammen, 1941.
41 Hecht, Wetzel, Behandlung, 1939, in: National Archives and Records Administration (USA), RG 238, PS-660.
42 Cf. Report of the deputy Gau Leader in the Warthegau, Ne/Wa., 1 Oct. 1941: concerning Midwife outreach in the Wartheland, in: IPN, GK 746/66, p. 105.
43 Report of the deputy Gau Leader in the Warthegau, Ne/Wa., 1 Oct. 1941: concerning Midwife outreach in the Wartheland, in: IPN, GK 746/66, p. 92; Letter of the health officer in Leslau, 6 Aug. 1940. In: Archiwum Państwowe w Włocławek (APW), 829/43.
44 With the introduction of the settlement permit regulated in the Reich Midwifery Law, midwives could be "transferred" to other parts of the country. Cf. Lisner, Hüterinnen, 2006.

were ethnic German resettlers.[45] At the same time, the number of Polish midwives fell from 800 to 600. Thus, obstetric care for German mothers by German midwives improved in relation to the German population, which increased up to 23% by 1944. In 1944, there were 3.4 German midwives per 10,000 Germans, but only 1.7 Polish midwives per 10,000 Poles.[46]

In October 1940, the health administration enacted the Reich Midwifery Law (Reichshebammengesetz) in the "incorporated eastern territories", which had been in force in the "Old Reich" since January 1, 1939, in order to regulate midwifery and to set ethnic bars straight according to the premises of racial segregation.[47] As in the "Old Reich", the law secured a monopoly for midwives in obstetrics by introducing the obligation to attend every birth and every miscarriage, thus strengthening their position as experts in pregnancy, childbirth and the postpartum period. The Reich Midwifery Law provided German midwives with extensive legal and economic protection. For example, it guaranteed them a minimum income, a pension and health insurance.[48] Polish midwives however, had no rights. They were only granted a licence, which was subject to "immediate revocation". This meant that German health authorities, such as public health officers, could ban Polish midwives from practising at any time with no possibility of appeal. As compulsory members of the Reich Midwifery Association, the Reich Midwifery Law placed Polish midwives under the supervision of the professional organisation, in addition to that of the public health officers. The head of the Reich Midwifery Association was Nanna Conti, the mother of the Reichsgesundheitsführer.[49] The district midwives, many of whom were Germans from the German Reich, were given extensive supervisory powers to monitor the quality of the professional practice of Polish midwives – a task that had been reserved for public health officers in the "Old Reich". In addition, Polish midwives had to pay 20% of their income to the professional organisation, which used the money to support German members. The Reich

---

45 Almost 200 of the midwives were Reich German. Cf. Allgemeine gesundheitliche Versorgung der Zivilbevölkerung (Public health care for civilians), 7 Aug. 1943. In: APP, 299 RSH/ 1882, p. 69–70.
46 Own calculation based on population statistics of 1944 (1.012.343 Germans and 3.40.000 Poles). Cf. Stiller, Politik, 2022, p. 1240.
47 Verordnung zur Einführung des Hebammengesetzes in den eingegliederten Ostgebieten vom 7. Oktober 1940, in: Reichsgesetzblatt I, 1940, p. 1333.
48 Cf. Tiedemann, Standesorganisation, 2001.
49 Cf. Peters, Biographie, 2018.

Midwifery Law restricted the activities of Jewish midwives to the care of Jewish women. The occupation administration declared that it was not responsible for the health and obstetric care of Jews and left the organisation to the Jewish communities.⁵⁰

About the "Evil" of Polish Midwives

Both health politicians and German midwives stereotypically described the quality of Polish midwives' work as poor. Margarete Riedel criticised that Polish midwives weren't familiar with postnatal visits and would attend the birth with little resources and effort, and then neither look after the mother nor the baby's navel.⁵¹

Just like Margarete Riedel, Nazi planners and health politicians regarded the admission of Polish midwives as an "evil" because of their training and professional practice.⁵² Above all, they rejected Polish midwives "as members of a foreign race", as a 1941 report by Schmalz's office put it. The report made it clear that because of their racial affiliation, Polish midwives did not provide the pregnancy counselling or postnatal education in the way that is "necessary in the interest of a healthy German offspring". "This educational work can only be done by German midwives", the report concluded.⁵³ The Schmalz office appears to have placed Polish midwives under general suspicion, assuming that they would counter the Germanisation and population policy objectives by making little effort to keep German mothers and children healthy.

In the Nazi midwives' journal, German midwives reported their shocking encounters with ethnic German traditions and thus emphasising the need to educate the ethnic German women. Midwife Ingeborg Morsbach, for example, told her colleagues in the "Old Reich" that ethnic German mothers – especially former Polish citizens – exclusively breastfed their babies for nine months or more without any supplementary porridge.⁵⁴ Her colleague Juli Prinz added

---

50   Cf. Hecht, Wetzel, Behandlung, 1939. In: National Archives and Records Administration (USA), RG 238, PS-660, p. 41.
51   Riedel, Eindrücke, 1941.
52   Cf. Report of the deputy Gau Leader in the Warthegau, Ne/Wa., 1 Oct. 1941: concerning Midwife outreach in the Wartheland, in: IPN, GK 746/66, p. 95.
53   Report of the deputy Gau Leader in the Warthegau, Ne/Wa., 1 Oct. 1941: concerning Midwife outreach in the Wartheland, in: IPN, GK 746/66, p. 95.
54   Cf. Morsbach, Erlebnisse, 1942.

that the mothers would let their babies crawl on the dirty floor and often used rags instead of nappies. She also criticised the fact that the woman giving birth was usually surrounded by a crowd of women, neighbours and relatives. Julie Prinz and Ingeborg Morsbach like German health experts, found these practices disturbing and dangerous to the health of both mother and child. As a result, Julie Prinz concluded, it was not surprising that only some of the ethnic German children survived their childhood.[55]

Health authorities in the Wartheland expected German midwives to intervene in traditional practices and also to discipline their clientele to practice comprehensive hygiene, such as washing their hands before each contact with the baby, and to feed the babies in a way that was considered healthy, which included breastfeeding at fixed times and for a fixed length of time.[56] Ethnic German mothers were to be modelled into NS-German and thus integrated into an "extended ethnic community" ("erweiterte Volksgemeinschaft") beyond the borders of the "Old Reich".[57] Belonging should be established and maintained through certain cultural practices and behaviours, as well as through a certain standardised approach and care of infants that is understood to be health-promoting.[58]

However, due to the shortage of German midwives until the end of the war, Polish midwives were an "evil" that could not be avoided, as the report by Schmalz's office concluded. More than one German health expert complained, that Polish midwives didn't intervene when they witnessed "health-threatening" practices among their clients.[59]

Despite the complaints of the Nazi authorities', Polish midwives and lay obstetricians were popular among the population. The Reich German midwife Julie Prinz reported: "I never heard any of the women who provided midwifery services being reprimanded; on the contrary, they were fanatically defended".[60] The involvement of a Polish midwife in a birth gave (ethnic) German women the opportunity to escape the control and discipline of a German

---

55　Cf. Prinz, Erlauschtes, 1943. See also Lisner, Hebammen, 2016.
56　Cf. Dr. Boening, Hauptamt für Volkswohlfahrt, Posen, 3 Feb. 1942: Anleitungen zur Säuglingsernährung. In: APP, 299 RSH/2073, p. 5–9.
57　Cf. Lisner (2016). On the concept of the "extended ethnic community" Cf. Kundrus, Regime, 2009.
58　Harvey, Women, 2003.
59　Report of the deputy Gau Leader in the Warthegau, Ne/Wa., 1 Oct. 1941: concerning Midwife outreach in the Wartheland, in: IPN, GK 746/66, p. 95.
60　Prinz, Erlauschtes, 1943, p. 202.

midwife from the Reich and to shape birth and the puerperium according to their own ideas and traditions. Due to the racist hierarchisation and disenfranchisement of the Polish population, it would hardly have been possible for Polish midwives to assume a position of disciplinary authority over their German clientele and to enforce Reich German standards. In addition, Polish midwives had long-standing contacts with the local Polish and ethnic German population, which enabled trust based relaionships. They were also familiar with local social norms and values. Reich German midwives, who had only lived in the Warthegau for a short time, did not have such contacts.[61]

In the district of Leslau, the public health officer tried to break down the relationships of trust between the population and the Polish midwives by means of controls. He forced the midwives to report separately all births in German families and to give reasons for them. The most common reason documented was "sudden birth", which the medical officer accepted in view of the poor transport conditions and the shortage of German midwives.[62] With the increasing presence of German midwives since 1941, however, these contacts came under scrutiny by the authorities.[63] Inspections by the Nazi Women's Office, ordered by the public health officer, show that German families who chose to use a Polish midwife despite the presence of a German midwife in the district were under surveillance.[64]

Conclusion

In line with the aim of Germanising the incorporated Polish territories and winning the declared "war of births" against the Poles, Nazi health authorities implemented a birth policy based on racial segregation. Thus, pregnancy and childbirth in the Warthegau meant very different things, depending on one's racial classification e.g. in the German Ethnic Register. The occupation regime supported German mothers financially and materially, and placed great emphasis on the health of both mother and child in order to increase the German

---

61   Cf. Lisner (2016).
62   Note of the health officer in Leslau, 10 May 1943. In: APW, 829/44.
63   German midwives were not distributed equally over the Wartheland. In some districts, mainly in the bigger cities, there were more German midwives than in the countryside. However, since 1941 their number continously raised. Cf.
64   Cf. Schreiben der NS-Frauenschaft, 3 Feb. 1944. In: APW, 829/41. See also Lisner, Hebammen, 2016.

population. Even more, they prioritised health resources for them. For example, midwives were instructed to give priority to German women.[65] The health of Polish mothers and children was only of interest in maintaining the Polish labour force. Polish women were only allowed a minimum period of maternity leave, and because all Poles were obliged to work from the age of 12, Polish mothers couldn't stay at home to take care of their children.[66] Abortions were encouraged for Polish women, some health experts even suggested not intervening if Polish women had miscarriages, and all resources such as midwifery, medicines, hospital treatment, but also baby cloths and feeding bottles were diverted to the German population.[67] The occupation administration denied any responsibility for the health and maternity care of Jews, leaving the organisation to the Jewish communities.[68] The Nazi birth policy practised in the "Old Reich", with its interdependent pronatalist and antinatalist orientation was transferred to the incorporated Polish territories, but adapted to the aims of the "war of births", which meant fighting the Poles "biologically". Health care and midwifery thus became instruments of racial segregation and Germanisation. At the same time, the assumption of being in a "war of births" with the Poles was reflected in the "Old Reich" and determined the treatment of Polish female forced labourers.[69]

In the annexed Polish territories, however, the aims of the Nazi birth policy clashed with the shortage of medical personnel in the incorporated Polish territories and the need for Poles as a work force. Although the number of German midwives increased by the end of the war, there were never enough to serve the entire German population, let alone the Polish. Particularly in rural areas,

---

65 Note of the Reich Minister of the Interior, 17 Jan. 1944, in: APP, 299 RSH/2070, Bl. 8.
66 Majer, Fremdvölkische, 1981, p. 373–385, 411. Albert Coulon: Bericht betr. Gegenwärtiger Stand der Polen-Politik im Reichsgau Wartheland im Rahmen der gesamten Polenfrage, 30 July 1941, in: Bundesarchiv Berlin (National Archive Berlin, BArch), R 138 II/2, p. 38.
67 Cf. Czarnowski (1996); Albert Coulon: Bericht betr. Gegenwärtiger Stand der Polen-Politik im Reichsgau Wartheland im Rahmen der gesamten Polenfrage, 30 July 1941, in: Bundesarchiv Berlin (National Archive Berlin, BArch), R 138 II/2, p. 38–42.
68 Cf. Lisner, Midwifery, 2020. See also: Wetzel/Hecht: Die Frage der Behandlung der Bevölkerung der ehemaligen polnischen Gebiete nach rassenpolitischen Gesichtspunkten, 25 Nov. 1939. In: National Archives and Records Administration (USA), RG 238, PS-660, p. 41.
69 Cf. Brüntrup, Verbrechen, 2019.

midwifery could not be provided without Polish midwives. Due to the decentralised organisation of midwifery care, birth attendance in women's homes provided scope for action that was not only geared to population policy but also to the wishes and needs of women and their families. The use of a Polish midwife offered German families the opportunity to escape the disciplinary grip of German midwives, who were mandated to educate their clientele and, in particular, to train them to obey German ideas about pregnancy, childbirth, postpartum and hygiene. Because of the racially inferior position of Polish midwives, they offered (ethnic) German mothers and families a somewhat greater room for manoeuvre and autonomy. Polish midwives as Poles subject to draconian punishment if they objected to Germans, had just few opportunities to educate their German clientele. Aware of this this situation, Nazi health authorities tightened their control over Polish midwives, although they were unable to dismiss them. German midwives together with public health officers, were entrusted with the supervision of Polish midwives and were thus able to increase their scope of action and power. In addition, members of German organisations such as the "Nationalsozialistische Volkswohlfahrt" (NSV, National Socialist People's Welfare) and the "NS-Frauenschaft" (NSF, National Socialist Women's League) were called upon to monitor encounters between Germans and Poles. In this way, the Nazi regime attempted to increase its hegemony over the female sphere of pregnancy, childbirth and postpartum and to invade the private sphere under occupation in a multi-ethnic environment.

As has been shown, questions of reproduction played a central role in the occupation and Germanisation policy in the incorporated Polish territories. An important question for further research would be the significance of birth policies and midwives in other countries occupied by Nazi Germany, such as Slovenia, where the German occupiers pursued a similarly radical Germanisation policy as in the Warthegau.[70]

Bibliography

Verordnung zur Einführung des Hebammengesetzes in den eingegliederten Ostgebieten. vom 7. Oktober 1940, in: Reichsgesetzblatt I 178, p. 1333.

---

70 Cf. Stiller, Politik, 2022.

ALBERTI, Michael, Die Verfolgung und Vernichtung der Juden im Reichsgau Wartheland 1939–1945, Wiesbaden, 2006.
ALBERTI, Michael, Exerzierplatz des Nationalsozialismus. Der Reichsgau Wartheland 1939–1941, in: Genesis des Genozids. Polen 1939–1941, Darmstadt, Klaus-Michael Mallmann/Bogdan Musiał, 2004, p. 111–126.
BASHFORD, Alison/LEVINE, Philippa (eds.), The Oxford Handbook of the History of Eugenics, 1. Aufl., Oxford University Press, 2010.
BRÜNTRUP, Marcel, Verbrechen und Erinnerung: das "Ausländerkinderpflegeheim" des Volkswagenwerks, Göttingen, Wallstein Verlag, Stadt Zeit Geschichte; Band 1, 2019.
BRZEZIŃSKI, Tadeusz (ed.), Historia medycyny, Warszawa, Państ. Zakład Wydawnictw Lekarskich, 1988.
CZARNOWSKI, Gabriele, Frauen als Mütter der Rasse. Abtreibungsverfolgung und Zwangseingriff im Nationalsozialismus, in: Unter anderen Umständen. Zur Geschichte der Abtreibung, Dortmund, Staupe, Gisela, 1996, p. 58–72.
EPSTEIN, Catherine, Model Nazi. Arthur Greiser and the occupation of Western Poland, Oxford, 2010.
ESCH, Michael, Gesunde Verhältnisse. Deutsche und polnische Bevölkerungspolitik in Ostmitteleuropa 1939 – 1950, Marburg, Marburg.
GROHMANN, Herbert, Erb- und Rassenpflege als Grundlagen biologischer Volkstumspolitik, 7 Oct. 1941. In: Archiwum Państwowe w Poznaniu (APP), 299 RSH/ 1137
HAAR, Ingo, Inklusion und Genozid. Raum- und Bevölkerungspolitik im besetzten Polen 1939 bis 1944, in: Deutschsein als Grenzerfahrung. Minderheitenpolitik in Europa zwischen 1914 und 1950, Essen, Beer, Mathias et. al., 2009, p. 35–59.
HANSSON, Nils/PETERS, Anja/TAMMIKSAAR, Erki: Sterilisierungsoperateur und Forscher: Leben und Karriere Benno Ottows (1884–1975). In: Medizinhistorisches Journal. Band 46, 2011, pp. 212–237,
HARVEY, Elizabeth, Harvey, Elizabeth (2003): Women and the Nazi East. Agents and Witnesses of Germanization, New Haven, 2003.
HECHT, Wetzel, Behandlung, 1939, in: National Archives and Records Administration (USA), RG 238, PS-660. I thank Alexa Stiller for letting me use the source.
HEINEMANN, Isabel, Rasse, Siedlung, deutsches Blut. Das Rasse- und Siedlungshauptamt der SS und die rassenpoltische Neuordnung Europas, Göttingen, 2003.

HERZOG, Dagmar, Unlearning Eugenics. Sexuality, Reproduction, and Disability in Post-Nazi Europe, Eisconsin, 2018.

KEHLE, Charlotte, Hebammen gesucht, in: Die Deutsche Hebamme, 56 (1941), p. 278–279.

KOCELA, Weronika & MALICKI, Jan, Trudna sztuka babienia: kultura medyczna polski drugiej połowy XVIII wieku, Diss., Katowice, 2017.

KÖNIG, Sophie, [D]ie Wacht an der wiege des deutschen Volkes. Leipziger Hebammen, ihre Betreuungsaufgaben und die Beteiligung an der nationalsozialistischen Bevölkerungspolitik, in: Die Rolle der Pflege in der NS-Zeit. Neue Perspektiven, Forschungen und Quellen, Stuttgart, Pfütsch, Pierre, 2024, p. 273–308.

KUNDRUS, Birthe, Regime der Differenz. Volkstumspolitische Inklusionen und Exklusionen im Warthegau und Generalgouvernement 1939–1944, in: Volksgemeinschaft. Neue Forschungen zur Gesellschaft des Nationalsozialismus, Frankfurt am Main, Frank Bajohr/Michael Wildt, 2009, p. 105–123.

KUŹMA-MARKOWSKA, Sylwia (1979-), Walka z "babkami" o zdrowie kobiet: medykalizacja przerywania ciąży w Polsce w latach pięćdziesiątych i sześćdziesiątych XX wieku, in: Polska 1944/45-1989: studia i materiały, 15 (2017), p. 189–215.

LINNE, Karsten, Struktur und Praxis der deutschen Arbeitsverwaltung im besetzten Polen und Serbien 1939–1944, in: Zwangsarbeit in Hitlers Europa. Besatzung, Arbeit, Folge, Brelin: Pohl, Dieter/ Sebta, Tanja, 2013, p. 39–62.

LISNER, Wiebke, Hebammen im Wartheland. Geburtshilfe zwischen Privatheit und Rassenpolitik, 1939–1945, Göttingen, 2025.

LISNER, Wiebke, 'A Birth Is Nothing Out of the Ordinary Here… ': Mothers, Midwives and the Private Sphere in the 'Reichsgau Wartheland', 1939–1945*, in: Private Life and Privacy in Nazi Germany, ed. Elizabeth HARVEY/Johannes HÜRTER/Maiken UMBACH/Andreas WIRSCHING, Cambridge University Press, 2019, p. 304–330.

LISNER, Wiebke, Hebammen im "Reichsgau Wartheland" 1939–45. Geburtshilfe im Spannungsfeld von Germanisierung, Biopolitik und individueller biographischer Umbruchsituation, in: Zwischen Geschlecht und Nation. Interdependenzen und Interaktionen in der multiethnischen Gesellschaft Polens im 19. und 20. Jahrhundert, Osnabrück, Matthias Barelkowski, Claudia Kraft, Isabel Röskau-Rydel (eds.), 2016.

LISNER, Wiebke, "Hüterinnen der Nation"? Hebammen im Nationalsozialismus, Frankfurt am Main, 2006.

MAJER, Dietmut, "Fremdvölkische" im Dritten Reich. Ein Beitrag zur nationalsozialistischen Rechtssetzung und Rechtspraxis in Verwaltung und Justiz unter besonderer Berücksichtigung der eingegliederten Ostgebiete und des Generalgouvernements, Boppard am Rhein, 1981.

MORSBACH, Ingeborg, Von der Arbeit in den Ostgebieten, in: Die Deutsche Hebamme, 57 (1942), p. 60–61.

MORSBACH, Ingeborg, Erlebnisse aus dem Warthegau, in: Die Deutsche Hebamme, 57 (1942), p. 172–173.

MUSIAL, Bogdan, Maren Röger, Kriegsbeziehungen. Intimität, Gewalt und Prostitution im besetzten Polen 1939 bis 1945. (Die Zeit des Nationalsozialismus.) Frankfurt am Main, S. Fischer 2015, in: Historische Zeitschrift, 304 (2017), p. 574–575.

PETERS, Anja, Nanna Conti (1881–1951): eine Biographie der Reichshebammenführerin, Berlin, 2018.

POHL, Dieter, Die Reichsgaue Danzig-Westpreußen und Wartheland: Koloniale Verwaltung oder Modell für die zukünftige Gauverwaltung?, in: Die NS-Gaue, ed. Jürgen JOHN/Horst MÖLLER/Thomas SCHAARSCHMIDT, De Gruyter Oldenbourg, 2007, p. 395–405.

PRINZ, J., Erlebtes und Erlauschtes einer Hebamme unter den Rücksiedlern im Wartheland, in: Die Deutsche Hebamme, 58 (1943), p. 202–204, 215–216.

RIEDEL, Margarete, Meine Eindrücke als Hebamme im Warthegau, in: Die Deutsche Hebamme, 56 (1941), p. 250.

RÖGER, Maren, Kriegsbeziehungen: Intimität, Gewalt und Prostitution im besetzten Polen 1939 bis 1945, Frankfurt am Main, S. Fischer, Die Zeit des Nationalsozialismus, 2015.

SCHMUHL, Hans-Walter, Das "Dritte Reich" als biopolitische Entwicklungsdiktatur. Zur inneren Logik der nationalsozialistischen Genozidpolitik, in: Tödliche Medizin. Rassenwahn im Nationalsozialismus, Kampmeyer, Margret (ed.) (2009).

STILLER, Alexa, Völkische Politik. Praktiken der Exklusion und Inklusion in polnischen, französischen und slowenischen Annexionsgebieten 1939–1945. 2 Teilbde. Göttingen, Wallstein 2022, in: Historische Zeitschrift, 319 (2024), p. 217–218.

TIEDEMANN, Kirsten, Hebammen im Dritten Reich: über die Standesorganisation für Hebammen und ihre Berufspolitik, Frankfurt am Main, Mabuse-Verlag, (Mabuse-Verlag Wissenschaft, Bd. 53), 2001.

Vossen, J., Extreme Typen – Die öffentlichen Gesundheitsdienste in Thüringen und im Warthegau im Vergleich, in: Das Gesundheitswesen, 75 (2013), p. 721–725.

Wetzel, Erhard & Hecht, Günther, Die Frage der Behandlung der Bevölkerung der ehemaligen polnischen Gebiete nach rassenpolitischen Gesichtspunkten (The question of the racial treatment of the population of the former Polish territories), 25 Nov. 1939. In: National Archives and Records Administration (USA), RG 238, PS-660. I thank Alexa Stiller for letting me use the source.

Wildt, Michael, Völkische Neuordnung Europas, in: Themenportal Europäische Geschichte, 2007, https://www.europa.clio-online.de/essay/id/fdae-1402. (05.06.2025).

Witte, Haupttagung der Hebammenschaft im Reichsgau Danzig-Westpreußen, in: Deutsche Hebamme, 55 (1940), p. 228–229.

Wolf, Gerhard, Ideologie und Herrschaftsrationalität: nationalsozialistische Germanisierungspolitik in Westpolen, Hamburg, Hamburger Edition, Studien zur Gewaltgeschichte des 20. Jahrhunderts, 2012.

Continuity of "Race Hygiene"?
Discourses and Practices of Sterilization in the Soviet Occupation Zone and the Early GDR

Stefan Jehne[1]

Abstract *This article addresses the question of why, despite the repeal of the Nazi sterilization law and an officially negative attitude towards eugenic sterilization, applications for sterilization were submitted and in some cases even carried out in the Soviet Occupation Zone and the early GDR. Stefan Jehne also examines the extent to which the sterilization policies and practices of the Soviet Occupation Zone and the GDR were influenced by their counterparts in the systems before.*

Introduction

With Command No. 6, published on January 8, 1946, the Soviet Military Administration in Germany abolished the Nazi sterilization law ("Gesetz zur Verhütung erbkranken Nachwuchses") in the Soviet Occupation Zone and declared it to be Nazi injustice.[2] Thus, forced eugenics legitimized by the state officially ended. As a consequence, several doctors and former lawyers who had participated in the implementation of the Nazi sterilization law, were prosecuted during the four years of the Soviet Occupation Zone. Yet only once, on March

---

1   Landeswohlfahrtsverband Hessen, Gedenkstätte Hadamar und Leibniz-Zentrum für Zeithistorische Forschung Potsdam.
2   See Command No. 6 of the Soviet Military Administration, January 6, 1946 (See for instance Bundesarchiv (BA) Berlin, DQ 1 Ministerium für Gesundheitswesen der DDR (MfG), No. 20992, sheets without pagination, letter from the Ministry of Labour and Health, allegedly signed by Maxim Zetkin, to the insurance company of Berlin, November 25, 1949.

27, 1946, were doctors legally convicted by the District Court in Cottbus.³ From the beginning of the prosecution, the relevant German health and judicial organizations were mostly opposed to the Soviet initiative to try those involved in sterilization practices during the Nazi era. For example, the president of the Central Department of Justice in the Soviet Occupation Zone, Eugen Schiffer, and the later first attorney general of the GDR, Ernst Melsheimer, argued after the Cottbus trial in March 1946 that prosecuting doctors for being involved in sterilizations would lead to a systematic exodus of medical doctors to the Western Occupation Zones. Because finally, according to Schiffer and Melsheimer, most doctors had been involved in national-socialist sterilizations.⁴

But these pragmatic motives were not the only reason why Schiffer, Melsheimer, and others intervened to stop the prosecutions. Indeed, the attitudes towards the legal evaluation of sterilizations in the Nazi-regime varied across relevant governmental organizations. For example, the leader of the Section II of the Central Department of Health, Werner Holling, argued that eugenically motivated sterilization during the Nazi period could not be illegal, because similar sterilizations were practiced in many other countries too.⁵ The same argument can also be found in the largely parallel West German debate, which differed structurally from the East German debate in that only in Bavaria, North Rhine-Westphalia and with cutbacks in Hesse was the "Gesetz zur Verhütung erbkranken Nachwuchses" repealed at all and was not qualified as a crime against humanity.⁶ Although the Soviets had classified the Nazi

---

3   See Ibid., sheets without pagination, judgement by the District Court in Cottbus against Ulrich Hammer, Paul Carthaus and Otto Bode, March 27, 1946. As we know today, the judgement was negotiated before between the Soviets and the German participants (See BA Stasi-Unterlagen-Archiv, MfS, BV Cbs, ASt 3441–54 VSG, p. 7–10, note by the Department of Justice of the Brandenburg Provincial Administration, signer not readable, March 15, 1946).

4   See Ibid., DP 1 Ministerium der Justiz der DDR (MdJ), No. 116, sheet 10, letter from the president of the Central Department of Justice in the Soviet Occupation Zone, Eugen Schiffer, to the Legal Department of the Soviet Military Administration in Berlin-Karlshorst, May 10, 1946; Cf. Meyer-Seitz: Verfolgung von NS-Straftaten, 1998, p. 54.

5   See BA Berlin, DQ 1 MfG, No. 20992, sheets without pagination, instructions concerning the penalization of sterilizations, no date. On the international eugenics movement and its transnational history of interdependence, See e.g. Allen, Eugenics, 2015, p. 224–232. Bashford, Internationalism, 2010, p. 43–61. Turda, Race, 2015, p. 62–79. Kühl, Rassisten, 2014.

6   See hierzu Tümmers, Anerkennungskämpfe, 2011, p. 43–45. Teicher, Mendelism, 2020, p. 205f.

Forced Sterilization Act as a Nazi injustice, the German efforts were successful. In May 1946, the Soviet Military Administration issued a new decree which defined only those Nazi sterilizations as crimes against humanity that were racially or politically motivated. Thus, eugenic sterilizations were no longer classified as crimes.[7]

Despite the abolition of the Nazi sterilization law by the Soviet Military Administration, medical doctors had internalized the administrative practices from the Nazi period and continued to request permission for sterilization. In the following, I would like to examine the sterilization debates and practices in the Soviet Occupation Zone and the GDR up to 1959 and ask whether this was something genuinely original, a continuation of Nazi racial hygiene or the reactivation of Weimar eugenics concepts. My particular focus is on the sterilization applications and partial executions with eugenic indications. In this regard, I ask how and why, despite a diametrically opposed official stance on eugenic sterilization applications and in some cases eugenic sterilizations were approved, even though there was no legal basis for this and any proactive action in this direction was therefore formally illegal. I also compare the sterilization debates practice in the Soviet Occupation Zone and early GDR with that in the Federal Republic of Germany during the same period, with the Nazi practice of forced sterilization and with that in the Weimar Republic in some extent. Finally, I will shed some light on the biographical influences of the actors involved.

The Renaissance of Weimar Eugenics? Sterilization Applications in the SBZ and GDR 1945-1959

Regarding the latter, two substantial differences in request practices before and after the end of the national-socialist regime 1945 can be easily identified. In the four years of the Soviet Occupation Zone, there were approximately 60 requests for sterilizations. Ca. five were provided with a eugenic indication and one of them had been approved.[8] During the 1950s, we can identify

---

7   See Meyer-Seitz, Verfolgung, 1998, p. 48.
8   Evidently, the not clearly definable dark figure of non-registered reports is not included, just as the actually executed sterilizations. Until the end of 1948, six to twelve requests per Department of health of each State had been registered (See BA Berlin, DP 1 MdJ, No. 7098, record note, signer not readable, December 29, 1948). Sächsisches

165 requests, approximately 85 of which were made for eugenic reasons. Ten of them had been certainly approved and further five were approved and executed.[9] In contrast, during the Nazi regime, about 400.000 sterilizations were carried out.[10] Thus, the number of requests for sterilizations and consequently the number of performed sterilizations, decreased dramatically after the Nazi era. This meant that the total number was significantly lower than in the western occupation zones, where 329 sterilization requests were submitted to the public health department in Bremen alone.[11] The figures were also significantly lower than in the Federal Republic of Germany in the 1950s. The head physician of the district hospital in Großburgwedel near Celle, Dohrn, who was indicted in 1962, had performed 162 sterilizations on women in 1958 and 1959 and, according to his own statements, a total of over 1000 since 1948.[12]

Also, most of these requests after 1945 were directed against women.[13] The reason for this is that official biopolitics in the Soviet Occupation Zone/GDR focused almost exclusively on the female body in matters of reproduction and contraception, as Daphne Hahn has already elaborated.[14] Another major difference is the fact that in all those requests it was claimed that the individuals submitted voluntarily. The conceptual background for the rejection of compulsory sterilizations but the approval of voluntary sterilizations reaches back to social democratic, socialist and bourgeois biopolitical programs in the Weimar Republic.[15] The Weimar sterilization debate had reached its climax with the draft law of the Prussian State Health Council from July 1932, which provided for voluntary eugenic sterilizations, the approval of which was to be decided by a panel of experts.[16] After 1945, there was a consensus across the occupation zones and subsequently in both German states that, in contrast to National

---

Hauptstaatsarchiv Dresden (SächsHStA DD), 11391 Landesregierung (LRS), Ministerium für Arbeit und Sozialfürsorge (MASF), No. 2144–2146.

9   See BA Berlin, DQ 1 MfG, No. 1843, Vol. II, No. 2036, No 2040, No. 6119, No. 7098 and Müller-Hegemann, Therapieversuche, 1958, p. 230–235.
10 See e.g. Baader, Eugenik, 2016, p. 319f.
11 See Nitschke, Erbpolizei, 1999, p. 267.
12 See Hahn, Modernisierung, 2000, p. 96.
13 In the 1950s, only 7 to 9 of 165 requests were directed against men; all others were directed against women. In one case the gender is not evident (See BA Berlin, DQ 1 MfG, No. 1843, Vol. II, No. 2040, No. 7098).
14 See Hahn, Modernisierung, 2000, p. 40–42, 207, 217–218, 228–230, 277–278, 305.
15 See Schwartz, Sozialtechnologien, 1995, p. 264–311.
16 See for instance Vossen, Umsetzung 2009, p. 98–100.

Socialism, only voluntary sterilizations should be carried out. This is exemplified by publications in this regard by the geneticist and Nazi perpetrator Hans Nachtsheim, who was at pains to emphasize the alleged abuse of eugenics by the Nazi regime.[17] However, Nachtsheim wanted to relativize the voluntary nature of people marked as "feeble-minded". Instead, the state should be able to order sterilization without consent in these cases.[18]

It was common during the four years of the Soviet Occupation Zone and the early GDR in the 1950s that medical doctors requested the sterilization of women together with a request for abortion. An example of a eugenic indication is the case of Irmgard P. in June 1949, who was described as "deaf and dumb". Allegedly, she had applied for her own sterilization at the Department of Health in Leipzig after her abortion had been approved. Her request for sterilization, which was signed by her but written in third person presumably by a doctor, was eugenically motivated according to today's and as well in the contemporary understanding. It was claimed that she was 'mentally below average', and that her deafness was inheritable.[19] The Department of Health in Leipzig forwarded the request to the Department of Health of the Saxon State, and its head, Friedrich Winkler, decided to authorize the sterilization without any legal basis.[20] Winkler can be seen as a paradigmatic example of individual continuities of eugenicists from the 1920s through National Socialist rule to the post-war period. From 1928 to 1934 he lectured at university of Rostock about eugenics and "race hygiene" and was a visiting scientist at the so-called Racial Biology Institute in Uppsala, Sweden. Afterwards, he worked as head of the Department of Health in Neustrelitz. During World War II he worked as a military hygienist at German *Wehrmacht*.[21] As I have already mentioned,

---

17   See Nachtsheim, Rassenmischung, 1947, p. 148–154. Nachtsheim, Sterilisierung 1952, p. 47–50. Also Tümmers, Anerkennungskämpfe, 2011, p. 51–54. and Doetz, Alltag, 2011, p. 213–215.

18   See Nachtsheim 1952, p. 47–50; in addition Tümmers, Anerkennungskämpfe, 2011, p. 54.

19   SächsHStA DD, 11391 LRS, MASF, No. 2146, sheet 81, self-application, signed by Irmgard P., filed at the Health Department of Leipzig, June 7, 1949.

20   Ibid., sheet 82, letter from the Department of Health of the Saxon state, signed by Friedrich Winkler, to the Department of Health in Leipzig, June 27, 1949.

21   See SächsHStA DD, 19117 Personalunterlagen sächsischer Behörden, Gerichte und Betriebe from 1945, Box 2097, Dossier of Friedrich Winkler, sheet without pagination, Dossier, handwritten by Friedrich Winkler, September 2, 1945; Ibid., DO 1 Ministerium des Innern (MdI), No. 102922 Dossier of Friedrich Winkler, sheet 1–3, ‚questionnaire', December 3, 1948, sheet 3, Dossier, selfwritten, no date; Ibid., R 4901 Reichsministeri-

legalizing formally voluntary sterilizations for eugenic reasons was discussed several times but officially rejected. Winkler thus concealed the real reason for the sterilization of Irmgard P. in his permission. Formally, he gave his consent for medical and social reasons. Three weeks later, the Central Department of Health finally passed a first official regulation concerning the sterilization complex of women. Henceforth, voluntary sterilizations of women for strictly medical indications were legal, but not for other possible indications.[22]

Debates and Legislative Initiatives in the Soviet Occupation Zone and Early GDR on the Legalization of Sterilization

During the Weimar Republic, the regions of Saxony and Thuringia in particular had been pioneers in both the discourse and the illegal practice of sterilization.[23] It is therefore not surprising that the first demand to regulate sterilizations by law, which was sent to the Central Department of Health of the Soviet Occupation Zone, was made by the Department of Health of the Saxon State in January 1948. In an internal letter, Wladimir Lindenberg, responsible in the Central Department of Health, stated he fully supported the Saxon request.[24] Together with the Central Department of Justice, the Central Department of Health tried to define a legal framework to outline the circumstances, in which sterilizations would be legal. Wladimir Lindenberg argued that sterilizing for medical reasons was indisputably legal, and he added that it should

---

um für Wissenschaft, Erziehung und Volksbildung (RMfWEV), No. 1320, index card No. 10539 (Friedrich Winkler); entry ‚Friedrich Winkler', Catalogus Professorum Rostochiensium, https://purl.uni-rostock.de/cpr/00001837 (05.06.2025).

22 This is comprehensible from the correspondence between the Saxon State Department of health and the Department of Health of the district of Leipzig (See Ibid., sheet 71, letter from the Saxon State Department of health and the Department of Health, signed by Schratz, to the Department of Health of the district of Leipzig, November 9, 1949).

23 For example in Zwickau, Saxony, the physicians Gustav Boeters and Heinrich Braun sterilized 67 people between 1921 and 1925 (See Braun, Sterilisierung, 1924, p. 104–106. Boeters, Unfruchtbarmachung 1925, p. 341.) In Thuringia it was Margarete Hielscher, who sterilized 27 people in 1924 to 1926 in the Psychiatric State Institute of Stadtroda, (See Hielscher, Schwachsinniger 1930, p. 97–99.)

24 See BA Berlin, DQ 2 Ministerium für Arbeit und Berufsausbildung der DDR (MAB), No. 3887, sheet 450, letter from Wladimir Lindenberg to Erwin Marcusson, January 10, 1948.

also be legal 'in all other cases'. These other cases were 'a clear, verifiable hereditary disease' (which points to eugenic motivations), the 'social situation', and 'the existence of a lebdinous psychopathic personality' (which points to psychiatric motivations).[25] At the end of January 1948, a meeting of section leaders took place within the Central Department of Health in Berlin. The head of the Section of Staff and Instructions, Carl Coutelle, stated that eugenically as well as socially motivated sterilizations could not be legalized. Only in cases of medical indications sterilizations shall be 'tolerated'.[26] Coutelle's position becomes clearer considering his biography. Already in the Weimar Republic he was a member of a Communist student group.[27] The communist party had regularly rejected eugenically motivated sterilizations, calling them 'class medicine'.[28]

It is astonishing that Wladimir Lindenberg reconciled his position with that of Coutelle. Apparently, a clear position on eugenically motivated sterilizations was not appropriate within the Central Department of Health. Their president, Karl Linser, instructed Lindenberg to prepare a bill regulating the sterilization topic in the following internal organizational meeting.[29] However it was not him, but employee Marie Schulte-Langforth, who prepared the pertinent bill in August 1948. She argued that sterilizations for medical reasons should be legal, but not for social reasons. But she further argued, one would have to respect a person's wish for sterilization if they were mentally or physically ill.

Although Schulte-Langforth was in favour of eugenics, the internal vote in the Central Department of Health on eugenically indicated sterilizations was clearly against it. Thus, she modified the bill, and the new version was much more restrained about eugenically motivated sterilizations.[30] However, her bill

---

25   Ibid. DP 1 MdJ, No. 7098, sheet 2, letter from Wladimir Lindenberg to the Central Department of Justice of the Soviet Occupation Zone, January 30, 1948.
26   Ibid. DQ 1 MfG, No. 20992, sheet without pagination, protocol of a meeting of section leaders within the Central Department of Health of the Soviet Occupation Zone, February 11, 1948.
27   See Ibid. DO 1 Ministerium des Innern (MdI), No. 94962, sheet without pagination, sheet of staff, no date.
28   See Benjamin, Klassenmedizin, 1925, p. 8–12. Ibid. Rassenhygiene, 1927, (quoted in Schwartz 1995, p. 80).
29   See BA Berlin, DQ 1 MfG, No. 20992, sheets without pagination, protocol of a meeting of section leaders within the Central Department of Health of the Soviet Occupation Zone, February 11, 1948.
30   Ibid. sheet 10, bill prepared by Marie Schulte-Langforth, October 4, 1948.

was not passed into law. In further discussions between Schulte-Langforth and the responsible staff of the Central Department of Justice, no agreement was reached.[31] Altogether, the various debates within and between the relevant organizations ended at the end of 1948 without a legal regulation. The informal practice of doctors registering people for sterilization continued and was not affected at all by these debates at the ministerial level.

The Central Administration for Public Health's negative position towards the legalization of eugenic sterilization did not change after the founding of the GDR on 7 October 1949. This, however, did not put an end to the attempts at sterilizing people for eugenic reasons.[32] This internalized need within the medical profession in the GDR led to the fact that the question of how to deal with eugenically and socially indicated sterilization requests was declared a top priority by the leading health official Maxim Zetkin in April 1950. The communist Zetkin wrote to the party executives of the SED that, on the one hand, he could – against the traditional communist position – certainly understand the desire to sterilize women because of the poor prospects for the heredity of their offspring, but on the other hand, he could also understand the hesitation of the central health department to explicitly regulate the issue by law. Despite his sympathy in principle for eugenically motivated sterilizations, Zetkin ultimately opposed their legalisation, solely for political reasons.[33] Regarding compulsory sterilizations, he made clear that it was politically impossible to reinstall eugenically motivated practices so fast after the recent Nazi sterilizations. Evidently, it was more important for the GDR to distance itself from the Nazi regime's extermination policy than to implement its own eugenics programmes, despite the fact that eugenics was still considered a serious biopolitical concept. There is no official response from the SED leadership to Zetkin's letter, but there is a handwritten note on this letter that the party's General

---

31   Ibid. DP 1 MdJ, No. 7098, sheet 13, recorded note by Marie Schulte-Langforth, December 29, 1948.
32   See for instance ebd, No. 2147, sheet 114, letter from Oskar and Charlotte T. via the medical department of Löbau to the Saxon State Department of Health in Dresden, December 16, 1949; sheet 115, "specialist medical report" concerning Charlotte T., prepared by Elfriede Ochsenfahrt, December 7, 1949. Charlotte T. was claimed to be schizophrenic (Ibid.).
33   See BA Berlin, DQ 2 MAB, No. 3887, sheet 433, letter from Maxim Zetkin, Ministry for Labour and Health to the SED party executives, April 15, 1950.

Secretary Walter Ulbricht personally rejected the request to formalise rules for sterilization. Again, no law was passed.³⁴

Despite this, health officials kept on demanding the creation of a sterilization law in the 1950s. For example, in November 1950, the head of the gynaecological clinic of the University of Rostock, Hans Hermann Schmid, and later in January 1958 his colleague from the Department for Social Hygiene, Karl-Heinz Mehlan, both called for a sterilization law based on eugenic ideas.³⁵ Schmid's request was directly supported by his colleagues of the gynaecological clinics of the universities at Jena (Gustav Döderlein), Leipzig (Robert Schröder), and Halle (Helmut Kraatz).³⁶ Apart of Schmid, the other three professors had been involved as perpetrators in carrying out forced sterilizations during the National Socialist era.³⁷ In his request, Schmid tried to construct a difference between eugenically motivated sterilizations of the national-socialists and sterilizations under democratic conditions.³⁸ The argument that eugenic steriliza-

---

34   See Ibid., handwritten note, no date and no author.
35   See SächsHStA DD, 11391 LRS, MASF, No. 2144, sheet 101, letter head of the gynaecological clinic of the university of Rostock, Hans Hermann Schmid, to the State government of Mecklenburg, Ministry of Health, November 21, 1950; BA Berlin, DQ1 MfG, No. 21170, sheet without pagination, letter from the head of the Institute for Social Hygiene at the university of Rostock, Karl-Heinz Mehlan, to the Head of the Main Department "Mother and Child" at the Ministry of Health of the GDR, Käthe Kern, January 21, 1958.
36   See SächsHStA DD, 11391 LRS, MASF, No. 2144, sheet 100, letter by the Head of the gynaecological clinic of the university of Leipzig to the Department of Health at the Saxon Ministry of Labour and Social Welfare, December 8, 1950.
37   Gustav Döderlein was assistant at the gynaecological clinic at the university of Berlin (until 1936) and afterwards he was head of the police state hospital in Berlin until the End of World War II. At both places he took through sterilizations. The same was true for Helmut Kraatz. He took through sterilizations at the gynaecological clinic at the university of Berlin until 1941. Robert Schroeder was already head of the gynaecological clinic at university of Leipzig during the national-socialist-period and took as well through sterilization operations even before 1933. (See BA Berlin, R 4931 RMfWEV, No. 13261, index card No. 1709 (Gustav Döderlein). Ibid. No. 13269, index card No. 5387 (Helmut Kraatz). Ebd, DQ 1, MfG, No. 24137 Dossier Helmut Kraatz, p.10-11, Dossier, April 24, 1953. Ibid. Reichsärzteregister, index card Robert Schröder. David, Döderlein, 2011, p. 196. Die Direktoren der Universitätsfrauenklinik p. 274; Doetz, Alltag, 2011, p. 213; Klee, Personenlexikon, 2003, p. 561; Klose, Nachuntersuchungen 1940 (simult. med. Diss. Kiel 1940), p. 4, 13.)
38   See SächsHStA DD, 11391 LRS, MASF, No. 2144, sheet 101, letter head of the gynaecological clinic of the university of Rostock, Hans Hermann Schmid, to the State government of Mecklenburg, Ministry of Health, November 21, 1950.

tions were a serious social technology that had been abused by the National Socialists was neither an exclusive argument of Hans Hermann Schmid nor a specific feature of the discourse of the Soviet Occupation Zone and early GDR. Rather, it was also the core argument of Hans Nachtsheim in his "Critique of National Socialist Racial Theory" or – in addition to the pragmatic reason of the shortage of doctors – the central political argument against the systematic prosecution of those involved in Nazi forced sterilizations in the Soviet Occupation Zone.[39] This dichotomous separation of eugenics from the Nazi practice of forced sterilization made it possible in the Federal Republic to continue offering eugenic counselling during the 1950s.[40]

However, the Central Department of Health of the GDR in Berlin did not comply with the scientist's demands: no law was passed to regulate this complex. On the contrary, the Central Department of Health passed another regulation in 1954 confirming the position of the first of 1949.[41] The same was true for the Case of Karl-Heinz Mehlan. In his request he talked about 'imbecile and psychopathic children, who will flood the recreation centres if they were not aborted before.' He further wrote that it was necessary to sterilize the mothers because 'unfortunately' there were 'some antisocial, moronic, and instinctive girls' who get pregnant all the time if they are not sterilized.[42] With attributes such as "asocial", "moronic" and "instinctive", eugenicists categorized actual or supposed (sexual) deviations from the norm with intersectional discriminating intent across all temporal and political system boundaries. This finding can also be applied to the eugenic sterilization practice in the Soviet Occupation Zone and the GDR in the 1950s, as can be exemplified paradigmatically by an application made in 1954 against Jutta L. from Leipzig for "feeblemindedness". There was no serious medical diagnosis here, but her supposed sexual deviation from the norm was medicalized. The psychiatrist who wrote the report described her sexual behaviour as "instinctive" and her in general as "antisocial" and "dull". A male doctor, and therefore a member of a privileged class, stigmatizes a member of a deprivileged class on the basis of his internalized

---

39 See Nachtsheim, Rassenmischung, 1947, p. 148.
40 See Vogel, Retinoblastom 1957, p. 565, 569. Schenk, Behinderung, 2016, p. 20f.
41 See Verfügungen und Mitteilungen des Ministeriums für Gesundheitswesen No. 2, March 16,1954, p. 6.
42 BA Berlin, DQ 1 MfG, No. 21170, sheet without pagination, letter from the head of the Institute for Social Hygiene at the university of Rostock, Karl-Heinz Mehlan, to the head of the Main Department "Mother and Child" at the Ministry of Health of the GDR, Käthe Kern, January 21, 1958.

classist and sexist system of prejudice and wants her to be sanctioned for deviating from the norm. He uses the stigma "antisocial" to express the supposed break with the class norm, denies her free will and control over her sexuality in a sexist way with the term "instinctive" and disparages her with the attribute "dull".[43]

Mehlan's demand was therefore by no means merely utopian but was already being applied in practice as a matter of course. However, his application was not successful but he did not have to bear any consequences – just as little as the doctors and Nazi perpetrators Robert Schröder and Johannes Suckow[44] who were involved in the sterilization application against Jutta L. So, the limits of the sayable and executable were much more extensive within the GDR's biopolitical organizational system than the official position of the Ministry of Health would suggest.

Conclusion

To sum up, the adoption of a new sterilization law was discussed in the Soviet Occupation Zone and in the early GDR but remained in some ways open-ended. Sterilization demands were practiced at least to the stage of making requests, but partially approved and executed. Hereby, the biopolitical focus was decidedly placed on the female body. So, there is no doubt that eugenic debates and practices continued after the end of World War II. Even during the persecution of doctors and lawyers involved in compulsory sterilizations, German officials were interested in rehabilitating eugenics as a serious biopolitical concept. But as I have pointed out, every request of sterilization was formally declared to be voluntary, and their total number was much lower than under National Socialism. Thus, both the discourse and the practice are more reminiscent of the Weimar Republic, although it cannot be ruled out

---

43   Ibid., No. 1843, Vol. II, sheet 61, Sending of Robert Schröder's application for sterilization against Jutta L. with an excerpt from the expert opinion of the senior physician and psychiatrist at the Leipzig-Dösen State Hospital, Johannes Suckow, by the head of the Mother and Child Department, Healthcare Division at the Leipzig District Council, Margarete Boenheim, to the Mother and Child Department at the Ministry of Healthcare of the GDR on 11 February 1954.

44   From December 1, 1942 to March 31, 1943, Johannes Suckow was in charge of research at the Wiesloch sanatorium and nursing home as part of the Nazi "euthanasia" program (See for instance Lienert, Euthanasie-Verbrecher 2018, p. 87–89).

that the actors involved in the Nazi practice of forced sterilization merely tactically adapted their sterilization applications and/or legal demands to the new discourse framework after 1945. However, on the level of the Central Department of Health and the later Ministry of Health of the GDR, the internal debates seemed partially to be a serious attempt to argue for a legalization of eugenically motivated sterilizations. But as Maxim Zetkin's letter to the SED party leadership made clear, the politically motivated demarcation from the abuse of sterilizations by the Nazi regime made it impossible for the GDR to officially install a biopolitical programme involving sterilizations for eugenic reasons.

Bibliography

BAADER, Gerhard, Eugenik, Rassenhygiene und "Euthanasie", in: FRIEDMAN, Alexander/HUDEMANN, Rainer (eds.), Diskriminiert – vernichtet – vergessen, Behinderte in der Sowjetunion, unter nationalsozialistischer Besatzung und im Ostblock 1917–1991, 2016, p. 311–320.

BASHFORD, Alison, Internationalism, Cosmopolitanism, and Eugenics, in: The Oxford Handbook of the History of Eugenics, ed. Alison BASHFORD, Philippa LEVINE, New York, Oxford University Press, 2010, p. 154–172.

BASHFORD, Alison/LEVINE, Philippa (eds.), The Oxford handbook of the history of eugenics, New York, Oxford University Press, Oxford handbooks, 2010.

BENJAMIN, Georg, Tod den Schwachen? Neue Tendenzen zur Klassenmedizin, 2. Aufl., Berlin 1925.

Boeters, Gustav, Die Unfruchtbarmachung Geisteskranker, Schwachsinniger und Verbrecher aus Anlage, in: Zeitschrift für Medizinalbeamte und Krankenhausärzte 38 (1925), S. 336–341.

BRAUN, Heinrich, Die künstliche Sterilisierung Schwachsinniger, in: Zentralblatt für Chirurgie 51 (1924), p. 104–106.

DAVID, Matthias, Gustav Döderlein. Von Stoeckel über das Berliner Polizeikrankenhaus nach Jena, in: Schleußner, Ekkehard (eds.): Vom Accouchierhaus zur Universitäts-Frauenklinik. 230 Jahre Frauenklinik Jena, Jena, Städtische Museen, 2011, p. 195–209.

Die Direktoren der Universitätsfrauenklinik Jena 1779–2008, in: SCHLEUßNER, Ekkehard (ed.): Vom Accouchierhaus zur Universitäts-Frauenklinik. 230 Jahre Frauenklinik Jena, Jena, Städische Museen, 2011, p. 267–276.

DOETZ, Susanne, Alltag und Praxis der Zwangssterilisation, Die Berliner Universitätsfrauenklinik unter Walter Stoeckel 1942 – 1944, Berlin, bebra, (Schriften-Reihe zur Medizin-Geschichte, Bd. 19) 2011.

HAHN, Daphne, Modernisierung und Biopolitik. Sterilisation und Schwangerschaftsabbruch in Deutschland nach 1945, Frankfurt, New York, Campus-Verl., Campus Forschung, 2000.

HIELSCHER, Margarete, Die Unfruchtbarmachung Schwachsinniger aus rassenhygienischen und sozialen Gründen (Univers. Diss.) Jena 1930.

KLEE, Ernst, Das Personenlexikon zum Dritten Reich. Wer war was vor und nach 1945?, Frankfurt/M., Fischer Taschenbuch, 2003.

Klose, Felicitas, Nachuntersuchungen des Schicksals der in den Jahren 1934 bis 1937 im Stadtkreis Kiel auf Grund des Gesetzes zur Verhütung erbkranken Nachwuchses sterilisierten Frauen unter Berücksichtigung der Frage nach der Notwendigkeit einer nachgehenden Fürsorge, Leipzig, G. Thieme, 1940, p. 4, 13.

KÜHL, Stefan, Die Internationale der Rassisten. Aufstieg und Niedergang der internationalen eugenischen Bewegung im 20. Jahrhundert, Frankfurt am Main, Campus Verlag, 2014.

LEVINE, Philippa, Anthropology, Colonialism, and Eugenics, in: Bashford, Alison/Levine, Philippa (eds.), The Oxford handbook of the history of eugenics, New York, Oxford University Press, Oxford handbooks, 2010, p. 43–61.

LIENERT, Marina, Johannes Suckow (1896–1994) – Ein »Euthanasie«-Verbrecher als Gründer der Klinik für Neurologie und Psychiatrie an der Medizinischen Akademie »Carl Gustav Carus« Dresden?, in: Kumbier, Ekkehardt/Steinberg, Holger (eds.): Psychiatrie in der DDR. Beiträge zur Geschichte (= Schriftenreihe zur Medizin-Geschichte, Bd. 24), Berlin 2018, p. 79–93.

MEYER-SEITZ, Christian, Die Verfolgung von NS-Straftaten in der sowjetischen Besatzungszone, Berlin, Berlin-Verl. Spitz, (Schriftenreihe Justizforschung und Rechtssoziologie) 1998.

MÜLLER-HEGEMANN, Dietfried, Über Therapieversuche mit Röntgenkastrationen bei multiple-Sklerose-kranken Frauen, in: Psychiatrie, Neurologie und medizinische Praxis. Zeitschrift für die gesamte Nervenheilkunde und Psychotherapie 10 (1958), p. 230–235.

NACHTSHEIM, Hans, Rassereinheit und Rassenmischung. Zur Kritik der nationalsozialistischen Rassentheorie, in: Das Deutsche Gesundheitswesen 2 (1947), p. 148–154.

NACHTSHEIM, Hans, Für und wider die Sterilisierung aus eugenischer Indikation, Stuttgart, Thieme, 1952.

NITSCHKE, Asmus, Die "Erbpolizei" im Nationalsozialismus, Zur Alltagsgeschichte der Gesundheitsämter im Dritten Reich. Das Beispiel Bremen, Opladen [u.a.], Westdt. Verl., 1999.

SCHENK, Britta-Marie, Behinderung verhindern. Humangenetische Beratungspraxis in der Bundesrepublik (1960er bis 1990er Jahre), Frankfurt/M. 2016.

SCHWARTZ, Michael, Sozialistische Eugenik, Eugenische Sozialtechnologien in Debatten und Politik der deutschen Sozialdemokratie 1890–1933, Bonn, Dietz, Reihe Politik- und Gesellschaftsgeschichte 42, 1995.

TEICHER, Amir, Social Mendelism. Genetics and the Politics of Race in Germany, 1900–1948, Cambridge, 2020.

TÜMMERS, Henning, Anerkennungskämpfe. die Nachgeschichte der nationalsozialistischen Zwangssterilisationen in der Bundesrepublik, Göttingen, Wallstein-Verl, Beiträge zur Geschichte des 20. Jahrhunderts, 2011.

TURDA, Marius, Race, Science, and Eugenics in the Twentieth Century, in, 2010, p. 62–79.

Vogel, Friedrich, Die eugenische Beratung beim Retinoblastom (Glioma retinae), in: Acta genetica et statistica medica 7 (1957), p. 565–572.

VOSSEN, Johannes, Die Umsetzung der Politik der Eugenik bzw. Rassenhygiene durch die öffentliche Gesundheitsverwaltung im Deutschen Reich (1923–1939), in: Wecker, Regina/Braunschweig, Sabine/Imboden, Gabriela/Küchenhoff, Bernhard/Richter, Hans Jakob (eds.), Wie nationalsozialistisch ist die Eugenik?×Internationale Debatten zur Geschichte der Eugenik im 20. Jahrhundert, Wien [u.a.], Böhlau, 2009, p. 93–106.

WECKER, Regina/BRAUNSCHWEIG, Sabine/IMBODEN, Gabriela/KÜCHENHOFF, Bernhard/RICHTER, Hans Jakob (eds.), Wie nationalsozialistisch ist die Eugenik?×Internationale Debatten zur Geschichte der Eugenik im 20. Jahrhundert, Wien [u.a.], Böhlau, 2009.

Physicians as the Main Actors in the Debate over Birth Control in Czechoslovakia, 1920s–1960s

Veronika Lacinová Najmanová[1]

Abstract *By the first half of the 20th century at the latest, birth control had become an integral part of public debate in many European countries, and the issue of contraception was discussed by various actors in relation to a broad range of political and social questions. Individuals and indeed entire organizations were driven to promote birth control by a variety of motives – most commonly based on medical, social, or eugenic considerations. In many countries, the women's movement also had a substantial influence, voicing its demands that women should be given the right to control their own bodies. This paper suggests that in Czechoslovakia, both during the interwar period and in the first two decades following the communist takeover, the medical aspect played the decisive role in motivating people to support or oppose contraception; and to a considerable extent medical considerations overshadowed other aspects of the debate (feminist, malthusian). As a consequence, physicians became the main actors in the debate over birth control.*

Keywords *contraception; medical discourse; reproductive rights; birth control; Czechoslovakia*

Introduction

Although until recently it seemed that the fight for reproductive rights had already been won in the last century in most countries of the Euro-American area, the development of recent years shows, on the contrary, that reproduction remains an important field on which cultural wars are waged and the right to control one's own reproduction continues to be endangered. At a time when

---

1   University of Pardubice, Faculty of Arts and Philosophy, Institute of Historical Sciences.

access to abortion or contraception is restricted in many countries and eugenically motivated interventions in reproduction are on the rise, more than ever, it seems relevant to follow the circumstances under which the struggle for reproductive rights was conducted in the past and to examine the discussions that took place around this topic.

The right to affordable and high-quality contraception began to be more clearly articulated in 19th century and became the main demand of the contraceptive movement, which developed in a number of countries no later than in the first half of 20th century. Birth control promoters sought to make contraception legal and widely available, and therefore spread awareness of the possibilities of birth control through lectures, leaflets and books, in some countries also fought for the legalization of contraception.[2]

From the 19th century onward, primarily under the influence of Robert Malthus,[3] deliberate birth control was discussed as a means of combating pauperism, and as one way of improving the situation of society's poor classes. In Malthus's view, the only acceptable means of birth control was the postponement of marriage or complete sexual abstinence. However, the proponents of neo-Malthusianism – a movement which began to emerge in an organized form on the national and international level around the turn of the 20th century[4] – considered various forms of contraception to be acceptable, and sought to promote them. It was also ca. 1900 that the eugenic aspects of birth control became increasingly emphasized – both from the perspective of its positive impact on reducing birthrates among so-called inferior persons, and also from the perspective of its negative impacts on fertility in healthy individuals of high eugenic quality.[5] In many European countries, the fears of overpopulation that had been voiced by Malthus in 1798 were replaced 100 years later by fears of a declining birthrate. As a result, the proponents of eugenics on the one hand emphasized the advantages of sterilization for certain groups of the population, while on the other hand they criticized the use of contraception by the middle classes; they stoked fears of the eventual

---

2   Jütte, Contraception, 2014, p. 106–174.
3   The English pastor's landmark work "An Essay on the Principle of Population" was first published in 1798, anonymously. The second revised edition was issued in 1803, this time under Malthus's name.
4   In 1877 the Malthusian League was established in England, followed by similar organizations in Germany, the Netherlands, Denmark, and other countries.
5   For more on the relationship between reproduction, gender, and eugenics, See e.g. Kline, Race, 2001 and Richardson, Love, 2003.

demise of the human race as a consequence of depopulation.⁶ During the first half of the 20th century, the issue of birth control became inextricably entwined with issues of women's rights. Movements promoting contraception targeted their efforts mainly at women, presenting birth control not only as a means of protecting women's health, but also as a way of liberating them from a frequently endless cycle of pregnancies and births.⁷

The debate over the acceptability and benefits of contraception encompassed numerous aspects, and it involved a wide range of actors from social reformers to clerics and politicians. From the outset of the debate, the medical perspective formed an integral and self-evident part of the discourse. It was usually physicians who invented new methods of contraception or improved traditional ones. It was also physicians who disseminated information about methods of birth control – either through direct contact with patients, or via various texts aimed at a broad general readership, primarily guides to married life or sexual health. For some types of birth control, the only (or almost only) way women could access contraceptives was to have them prescribed or applied by doctors.⁸ If we consider this in combination with the fact that from at least the 19th century, physicians became increasingly respected authority figures in society as a whole (becoming important partners of the state in cultivating a healthy and strong nation),⁹ then it is hardly surprising that in most countries they played a key role in issues of birth control.

Although the importance of doctors in the promotion of contraception has already been reflected by a number of researchers,¹⁰ doctors are often perceived in connection with reproductive rights mainly as a conservative group

---

6   For more on the relationship between reproductive policy and fear of depopulation, see e.g. Usborn, Body, 1992, p. 1. Szabó, Potraty, 2020, p. 33.
7   See Gordon, Body, 1976. as well as Grossmann, Reforming sex, 1995.
8   Primarily cervical caps, which during the first half of the 20th century became one of the most widely recommended (and mainly in Western countries, also used) methods of birth control. They were available on the basis of a medical prescription, and physicians also instructed women on how to fit them. Likewise, the fitting of intrauterine devices (IUDs) requires the assistance of doctors.
9   In the spirit of the processes described by M. Foucault in terms of medicalization and biopower. See Foucault, Discipline, 1977. Foucault, Historie, 1984. Foucault, Biopolitics, 2008.
10 Gawin, Interwar Poland, 2008. Kelly, Pill, 2020. Olszynsko-Gryn und Rusterholz, Politics, 2019. Rusterholz, Women Doctors, 2019.

that rather opposed their promotion.[11] I will try to relativize this opinion. The aim of this paper is to emphasize the importance of the medical aspect in discussions about birth control and in its promotion, using the example of the involvement of physicians in the promotion of contraception in Czechoslovakia in the period from the 1920s to the 1960s. On the following pages I will try to increase awareness of the role of physicians in the promotion of contraception not only in the interwar period but also in the post-war period, to point to a certain continuity in the approach to contraception in the two different political regimes, and to stimulate a discussion about the promotion of contraception in socialist countries. I will try to briefly outline that physicians played a vital role in the discussions about contraception both in the interwar period and in the 1950s and 1960s, and the medical aspect of family planning in Czechoslovakia strongly overshadowed other motivations associated with contraception (including neo-Malthusian or feminist motives). The reasons for the predominance of the medical view of contraception cannot be discussed in a paper of this scope, so in the following pages I will focus only on sketches of how physicians dominated discussions on contraception, first in interwar Czechoslovakia, when the topic of contraception had just been established in medicine, secondly in the 1950s and 1960s, when, on the contrary, the promotion of contraception received state support.

The presented partial conclusions are results of a larger research project investigating the promotion of contraception in Czechoslovakia between 1900 and 1970. Conclusions are based on the analysis of the Czechoslovak medical discourse (medical articles and books published in the examined period) and on the research of archival sources mapping the activities of selected physicians, birth control associations and the selected state administration authorities.

Physicians and Birth Control in Interwar Czechoslovakia

In Czechoslovakia physicians were the main actors not only in general debates concerning reproductive policy and reproductive health, but also in direct dis-

---

11   Ignaciuk et. Al. Doctors, 2014. Shmidt, Embodiment, 2018. Dudová, Framing, 2010.

cussions on the importance or effectiveness of contraceptive methods.[12] There was no mass movement that would seek to promote contraception and emphasize its social or "feminist" significance, therefore the health aspect of contraception was the one that significantly prevailed in social discourse.[13] The first substantial signs of interest in the issue of birth control in Czechoslovakia appeared around 1900, but up until the First World War contraception was to a large extent a taboo subject,[14] even within the medical profession; most medical practitioners considered sexual abstinence to be the only correct way of regulating the size of families.[15] Increased interest in birth control and related issues began to emerge in Czechoslovakia during the 1920s, and especially in the 1930s in connection with discussions regarding the decriminalization of abortions. The high rate of illegal abortions in interwar Czechoslovakia led to at least five proposals to legalize such terminations or to expand the range of circumstances in which an abortion could legally be carried out.[16]

These attempts sparked a wide-ranging debate not only on the political scene, but naturally also among experts in various fields, including the medical profession. Physicians were strongly opposed to expanding the range of permissible circumstances to include criteria motivated by social and eugenic concerns; they pointed out that their profession was about saving lives, not taking lives. They also argued that even abortions carried out in proper medical

---

12   Regarding the debate in Poland see Marcin Wilk, From Girls into Women, from Boys into Men, as well as Elisa-Maria Hiemer, Divergent Narratives on Family Planning, and Małgorzata Radkiewicz, Single mothers and the issue of motherhood in this volume.
13   For more on the importance of health motives in the Czechoslovak birth control movement, see Lacinová Najmanová, Health, 2021, p. 320–324.
14   An exception is the condom; condoms were the subject of much discussion due to the war, as they were viewed as a way of protecting soldiers from venereal diseases, though their use as a method of birth control was usually either criticized or not mentioned.
15   For example, the doctor Karel Malý in his book "Žena její krása a život pohlavní" ("Women, their beauty and sexual lives") states: "The most natural, correct and also moral way of preventing pregnancy would be complete abstinence from sexual intercourse. It has been proved that healthy and rational people can live for years without this intercourse." Karel, Žena, 1920, p. 111.
16   Up to this point, Czechoslovak law still contained provisions inherited from the legislative framework of the defunct Austro-Hungarian Monarchy; section 144 of the Criminal Code defined abortions as a criminal offence carrying a custodial sentence of 5–10 years. The only exception was in cases when the mother's life was at risk; in such cases abortions were legally permissible.

facilities represented a substantial risk to women's health and lives.[17] In fact, the law in interwar Czechoslovakia was not changed to expand the range of circumstances in which abortions were permissible, though they did nevertheless have an effect on how issues of birth control were perceived. They led many physicians (and other opponents of abortions) to re-evaluate their stance on contraception, which had previously been rejected mainly due to its unreliability as well as due to the perceived threat of depopulation.

As a result, during the interwar years the topic of birth control gradually found its way into various publications aimed at the general public as well as into specialist medical journals. There was a growing market for instructional publications on married life and sexual health, which treated birth control as an integral part of the wider topics of sexuality and marriage. A noticeable shift can also be observed in specialist medical journals; initially they mainly reported on research and publications from other countries, but later there was a shift towards Czechoslovak research as well as critical evaluations of individual methods of contraception.[18]

One of the key figures among the promoters of birth control in the interwar period was doctor Antonín Ostrčil – though his stance on the artificial regulation of fertility was in fact quite ambivalent. Ostrčil was one of the first Czechoslovak physicians to take a professional interest in birth control. In the 1930s, he headed a team studying the effectiveness of the calendar method of birth control, involving the calculation of (in)fertile days in the menstrual cycle; this method was widely discussed at the time, and Ostrčil became one of its most prominent critics. In a gynaecology textbook published in 1933,[19] he devoted several pages to birth control – not only presenting specific methods, but also discussing in which circumstances it was appropriate for a physician to prescribe or explicitly recommend contraception. He emphasized the essential importance of using contraception in cases of severe medical problems (a stance that was widely accepted at the time); he also considered contraception acceptable for social or eugenic reasons.[20] However, he went further than this, expressing the view that contraception was also acceptable for purely personal reasons; this view placed him at odds with the large majority of gynaecologists,

---

17   Pelcl, Stanovisko lékařské, 1930, p. 288–291.
18   The first scientific studies of contraception in Czechoslovakia were not conducted until the 1930s (see below). See Ostrčil, Poznámky, 1938, p. 205–207.
19   Ostrčil, Klinická gynekologie, 1933.
20   Ostrčil, Klinická gynekologie, 1933, p. 472–473.

who at the time were only prepared to tolerate contraception for medical reasons (or, in exceptional and severe cases, for social reasons). Although Ostrčil saw contraception as a means of preventing abortions, like many of his colleagues, he too was distrustful of it – mainly due to the risk it posed in the context of a declining birthrate. For this reason, Ostrčil was highly critical of attempts to disseminate information about birth control among the general public. Although he eventually allied himself with an organization that promoted birth control (see below), he nevertheless fundamentally believed that the pro-birth control movement represented a negative development.[21] He therefore supported contraception, but under clearly defined circumstances and, above all, under medical supervision.

While the medical community was slowly and cautiously beginning to accept birth control as part of its scope of interest, the women's movement in Czechoslovakia at the time did not yet pay much attention to this topic. In neighbouring Germany, for example, birth control was promoted by left-wing women's organizations;[22] in Czechoslovakia, however, the women's movement adhered to liberal ideology and birth control was not a central issue for them,[23] – moreover, their stance on abortion was inconsistent and hesitant.[24] Nevertheless, attempts to deal with the problem of illegal abortions (and their neg-

---

21 At a meeting of the Czech Medical Association (Spolek lékařů českých), he made the following statement: "I was invited to the inaugural session of this association at our clinic in order to express a gynecologist's opinion on this matter. I did so very willingly, because I considered it to be my duty, if it is no longer possible to rid ourselves of this new cultural spectre, then at least to banish it to the sidelines." Ostrčil also informed the Czech Medical Association that he considered contraception to be a double-edged sword, stating that if the only aim of the Association for Birth Control had been to reduce the birthrate, the state would not have permitted it to be established. Časopis lékařů českých, 1934, no. 39, p. 816–817.
22 See Usborn, Body, 1992. Grossmann, Reforming sex, 1995.
23 Left-wing women's organizations reflected more on the topic, but due to significant differences of opinion, these associations were not part of the Women's National Council – an umbrella organization for all the women's organizations in Czechoslovakia.
24 The Women's National Council was also asked to comment on the proposed reforms to section 144. The individual organizations that had the opportunity to express an opinion did not adopt a consistent stance; however, in general terms they opposed the inclusion of social circumstances among the criteria for an abortion to be permissible, and some organizations (mainly Catholic groups and associations of midwives) rejected any changes to the existing legislative provisions. See NA, Fonds Ženská národní rada, box 24.

ative impacts on the lives and health of thousands of women) led to the establishment of two organizations which promoted birth control. In 1932, an association entitled *Medical Protection for Women* (Zdravotní ochrana ženy) was founded in Brno. Its declared aim was to reduce the number of illegal abortions and to set up the first birth control counselling centre in Czechoslovakia. Two years later, in 1934, the *Association for Birth Control* (Svaz pro kontrolu porodů) was set up in Prague, with the same aims.

Although they were women's organizations, the topic of contraception was conceived primarily in terms of its importance for women's health, and therefore the health of society as a whole; thus, the motive for liberation or the emancipation of women was missing here. The influence of physicians (and the importance of medical aspects of birth control) is evident in both organizations – not only in the manner in which they presented their purpose and activities, but also in the links between the Prague association and a particular medical institution; the association openly declared its subordination to the authority of the head physician of a gynecology clinic. The Brno association's declared aims included its desire for contraception to become a part of public health care and a subject of systematic medical research.[25] From the very outset, the Prague association's counselling centre was linked with the above-mentioned doctor Antonín Ostrčil, who was the head physician at Charles University's gynaecology clinic no. 2 during the 1920s and 30s; the counselling centre was opened at the clinic in 1935. If we observe how the establishment of the association was presented in the press, it is evident that the emphasis is on the fact that the establishment took place only after consultation with medical circles, and also that the association clearly declares in its activities subordinate to the authority of Ostrčil.[26]

The State's Interest in Expanding Contraception in the 1950s and 60s

In the interwar years, Czechoslovak physicians took a reserved stance on birth control, which only emerged very gradually as a serious subject of their professional interest, publications and educational activities. Even the war years did not favor a more massive boom in interest in contraception. A rather pronatalist approach was on the rise, influenced first by the fear of a national threat ex-

---

25   MZA, Fonds Zemský úřad Brno, box 2936, reference no. 44268.
26   MUDr. M. N., Omezení porodnosti, 1935, p. 15.

acerbated by the rise of Nazism and then by the decline in the birth rate caused by war losses.[27] Nevertheless, after the Second World War (when the communist regime introduced a new public health care system), birth control quickly became accepted as an important subject for scientific study as well as a crucial element in creating "the socialist family" and physicians started playing an increasingly influential role in promoting birth control. As had been the case before the war, interest in contraception rose particularly in the context of the decriminalization of abortions. Czechoslovakia legalized abortion in 1957. Previously, the only permissible grounds for carrying out an abortion had been health-related, but from 1957 the list of criteria was expanded to incorporate economic and social circumstances.[28] Any woman, regardless of her marital status, could request an abortion if she was able to prove to a so-called abortion committee that she met these legal criteria.[29] However, there were fears of a rapid rise in the number of abortions, so before the new legislation came into force, the Ministry of Health took steps to promote contraception and make it more widely available; this was again considered a preferable means of controlling family sizes.[30]

In 1956, before the new legislation was passed, a birth control counselling centre was set up at Charles University's obstetrics and gynaecology clinic no. 3. The centre was headed by the doctor Ladislav Hnátek. It was intended to serve as a model for establishing other similar counselling centres in major cities, ones that would provide birth control advice to the largest possible number of people. The new counselling centres had several aims. Primarily, they

---

27   Rákosník/Šustrová, Population, 2018, p. 181–183.
28   Besides health-related reasons, the law also set out a number of other reasons which were classified as being "worthy of special consideration". These applied to women over the age of 40 and/or with at least three children, difficult circumstances experienced by unmarried pregnant women, family breakdowns, or potential negative impacts on living standards. For more information see Dudová, Interrupce, 2012, as well as Dudová, Framing, 2010.
29   It is worthy of note that before the legislation was amended in 1962, the abortion committees consisted of two medical professionals plus one woman appointed by the Communist Party's local district organization. After the amendment, the committee comprised one doctor.
30   Part of the law decriminalizing abortion was Directive No. 68, which established the obligation of doctors to deal with contraception as part of prevention in the care of a woman. Houdek, Historický, 1969, p. 110.

were to serve as places where members of the public would have access to verified information on birth control options (or to contraceptives themselves). Another important purpose of the centres was to train gynaecologists in issues related to birth control, so that all women would be able to consult their gynaecologist and receive reliable information on current options. Finally, the centres were to conduct research – studying the quality of existing birth control methods, improving these methods, and developing their own contraceptives so that Czechoslovakia would no longer be reliant on imports.[31] Research was also to focus on how contraceptives were used in practice.[32]

During this period, contraception was viewed in Czechoslovakia as an important aspect of preventive medicine. In countries with a growing population, the intention was to use birth control as a tool to combat the threat of overpopulation. But this was not the case in Czechoslovakia, where there was considerable reticence towards Malthus's theory, among other things, for ideological reasons.[33] By contrast, in Czechoslovakia as well as in the other states of Eastern bloc[34] it was presented as an important resource for preventing abortions, and it was thus promoted as a way of improving women's reproductive health.[35] Besides the desire to reduce the number of gynaecological problems caused by inexpertly conducted abortions (or indeed all abortions), attitudes to contraception in Czechoslovakia were also shaped by the desire to protect women's sexual health – especially to prevent the occurrence of certain pathological sexual conditions (such as a lack of sexual desire in women) and thus to

---

31   ZM, Fonds Ladislav Hnátek, manuscript of a lecture given at a conference abroad.
32   Research into the use of contraception and related issues of parenthood and sexuality was already carried out in 1956 by the State Statistical Office (Srb/Kučera, Vysušilová, 1959). In late 1958 and early 1959 the State Population Committee conducted a study in conjunction with the Ministry of Health (Srb/Kučera, Vysušilová, 1959.).
33   Malthus's population theory was already rejected in the 19th century by Karl Marx and Friedrich Engels, among other reasons, because it did not take into account the relationship between social capital and labor power. They also criticized him for seeking to maintain a social status quo that suits the ruling class. In the second half of the 20th century, a number of Eastern Bloc countries therefore adopted a rather "anti-Malthusian" approach. Olšáková/Janáč, Kult jednoty, 2018, p. 86.
34   As such e.g. Ignaciuk, Contraception, 2020, p. 1.
35   According to some studies at the time, up to a third of gynecological problems (e.g. infertility or even cancer) occurred as a consequence of abortions. See ZM, Fonds Ladislav Hnátek, manuscript of a lecture given at a conference abroad.

prevent marital problems would ultimately lead to family breakdowns.[36] Many studies cited by those in favour of contraception claimed that the fear of pregnancy was a frequent reason why women feared sexual intercourse – a situation which had a negative impact on spouses' intimate relations and could sometimes lead to divorce.[37] For this reason, contraception was paradoxically considered to be a potential means of increasing the birthrate, as it enabled couples to enjoy sexual intercourse without having to fear pregnancy, so they could plan to have children when it best suited them.[38]

Physicians played an important role in the promotion of contraception in Czechoslovakia in 1950s and 1960s for another reason as well. Given that the feminist movement was virtually eliminated here after the communist coup, it was physicians who took a leading role in articulating reproductive rights issues. While in a number of Western countries reproductive rights have been articulated through the women's movement or in general through civil activism, in the Eastern bloc, these themes were often articulated by physicians who have largely adopted a pro-women's agenda. As Hana Havelková points out, professionals, including physicians, due to the growth of technocracy, even under the conditions of a socialist society, had a certain degree of freedom and functioned as certain 'mediators' between political goals and people's everyday experiences.[39] In the field of contraception promotion or, more broadly, in the articulation of reproductive rights, this thesis of Hana Havelková proves to be valid.[40]

---

36 It is likely that this was closely connected with the development of sexology, with its emphasis on women's sexual pleasure (as has been pointed out by Lišková, Liberation, 2018.
37 According to Hnátek, the fear of pregnancy was very common; data from his counselling centre indicated that almost 60% of women feared that they would become pregnant after sex, and in 9% of cases this fear reached such a level of intensity that it led to a lack of sexual desire (frigidity). He stated that this situation could be prevented by the appropriate use of contraceptives; if women could be persuaded to use appropriate contraceptives, their fear of becoming pregnant would subside, and this could help save their marriages.
38 For example, Hnátek notes that "the correct contraception is not only an important tool against abortions, it is also one means of preventing sexual disorders in women." ZM, Fonds Ladislav Hnátek, Note on a lecture given by Professor J. Hynie on the causes of sexual disorders in women.
39 Havelková, Genderová politika, 2015, p. 160–163.
40 For the Polish debate during the final communist decade see Michael Zok, Killing of Unborn Children in this volume.

Conclusion

Physicians stood at the forefront of efforts to promote birth control in Czechoslovakia. Although their attitudes towards contraception changed and their motives varied from period to period, physicians appear to be fundamental factors in the promotion of contraception. Physicians not only conducted research on contraceptive methods, but also published books for the general public and actively advocated the promotion of contraception, either through the help of women's associations during the interwar Czechoslovakia or as part of state-organized health care in socialist Czechoslovakia. Not only due to the privileged social status of physicians, but also due to the indifference of the women's movement to reproduction, health motives and emphasis on reproductive health of society were dominant in the discussion of contraception and especially in the interwar period overshadowed other motives.

The aim of this case study was not only to report on the role of physicians in the contraceptive movement in one of the Central European countries, but above all to stimulate new ways of thinking about the role of physicians in the promotion of reproductive rights. The topic of the influence of expert discourse on the articulation of reproductive rights is an extremely current topic and deserves a more detailed evaluation. Although we know a lot about the ways in which physicians entered the field of birth control in the past, there are still many unanswered questions. These are, for example, mutual contacts between physicians from different countries, especially contacts across the Iron Curtain or the question of continuity or discontinuity of medical narratives related to reproductive freedom in the period before and after the Second World War.

Archival Sources

Národní archiv (NA) [National Archive], Fonds Ženská národní rada, box no. 24.
Moravský zemský archiv (MZA) [Moravian Provincial Archive], Fonds Zemský úřad Brno, box 2936, reference no. 44268.
Zdravotnické muzeum (ZM) [Medical Museum], Fonds Ladislav Hnátek, rukopis přednášky na zahraniční konferenci [manuscript of a lecture given at a conference abroad].
Zdravotnické muzeum (ZM) [Medical Museum], Fonds Ladislav Hnátek, poznámky k přednášce prof. Dr. J. Hynieho o příčinách sexuálních poruch

ženy [note on a lecture given by Professor J. Hynie on the causes of sexual disorders in women].

Bibliography

Primary sources

HOUDEK, J., Polák, Z. Historický přehled antikoncepce [Historical overview of contraception]. In Československá gynekologie 1969, 34(1-2): 110.
MALÝ, Karel. Žena její krása a život pohlavní [Women, their beauty an sexual lives]. Praha, 1920.
MUDr. M. N. "Omezení porodnosti" [Birth rate restrictions]. Rozkvět – Česká selka obrázkový týdeník, no. 43 (1935): 15.
OSTRČIL, Antonín. Poznámky k otázce o fysiologické plodnosti a neplodnosti ženy [Notes on the question of physiological fertility and infertility of women]. In Časopis lékařů českých 1938, roč. 18, č. 9, s.205-207.
OSTRČIL, A. Klinická gynekologie pro lékaře a mediky [Clinical gynecology for doctors and medics]. Praha, 1933.
PELCL, Hynek. "Stanovisko lékařské a sociálně zdravotní k otázce umělého přerušení těhotenství a ochraně před početím" [The opinion of medical and social health on the issue of abortion and contraception"]. Praktický lékař 10, no. 8 (1930): 288–91.
SRB, Vladimír, KUČERA, Milan. Výzkum o rodičovství 1956 [Research on parenting 1956]. Praha, 1959.
SRB, Vladimír, KUČERA, Milan, Vysušilová. Průzkum manželství, antikoncepce a potratů [Survey of marriage, contraception and abortion]. 1959, unpublished

Secondary literature

DUDOVÁ, Radka. Interrupce v České republice: zápas o ženská těla [Abortion in the Czech Republic: the struggle for women's bodies]. Praha: Sociologický ústav AV ČR, 2012.
DUDOVÁ, Radka, The Framing of Abortion in the Czech Republic: How the Continuity of Discourse Prevents Institutional Change. In Sociologický časopis / Czech Sociological Review 2010, 46(6): 945–976.

FOUCAULT, Michel, Discipline and Punish: The Birth of the Prison. London: Allen Lane, 1977.

FOUCAULT, Michel, Histoire de la sexualité. 1, La Volonté de savoir. Paris: Gallimard, 1984.

FOUCAULT, Michel, The Birth of Biopolitics: Lectures at the Collège de France, 1978–79. Basingstoke: Palgrave Macmillan, 2008.

GAWIN, Magdalena, The sex reform movement and eugenics in interwar Poland. In Studies in History and Philosophy of Biological and Biomedical Sciences 2008, 39: 181–186.

GORDON, Linda, Woman's Body, Woman's Right: A Social History of Birth Control in America. New York: Grossman Publishers, 1976.

GROSSMANN, Atina. Reforming sex: The German Movement for Birth Control and Abortion Reform, 1920–1950. New York, NY: Oxford University Press, 1995.

HAVELKOVÁ, Hana. (De)centralizovaná genderová politika: Role Státní populační komise [(De)centralized gender policy: The role of the State Population Commission]. In Havelková, Hana, Oates-Induchová, Libora (eds). Vyvlastněný hlas: proměny genderové kultury české společnosti 1948–1989 [The dispossessed voice: changes in the gender culture of Czech society 1948–1989], Praha: Sociologické nakladatelství, 2015, 125–168.

IGNACIUK, Agata, No Man's Land? Gendering Contraception in Family Planning Advice Literature in State-Socialist Poland (1950s–1980s). In Social History of Medicine 2020, 33 (4): 1327–1349.

IGNACIUK, Agata, ORTIZ-GÓMEZ, Teresa, RODRÍGUEZ-OCANA, Esteban, Doctors, women and circulation of knowledge on oral contraceptives in Spain: 1960s–1970s. In Ortiz-Gómez, Teresa and Santesmases, María Jesús (eds). Gendered drugs and Medicine. Historical and sociocultural perspectives. London: Routledge, 2014,

JÜTTE, Robert, Contraception: a history. Cambridge: Polity Press, 2008.

KELLY, Laura, The Contraceptive Pill in Ireland c.1964–79: Activism, Women and Patient Doctor Relationships. In Medical History 64 (2020), p. 195–218.

KLINE, Wendy. Building a Better Race: Gender, Sexuality, and Eugenics from the Turn of the Century to the Baby Boom. Berkeley, CA: University of California Press, 2001.

LACINOVÁ Najmanová, Veronika. Reproduction between Health and Sickness: Doctors' Attitudes to Reproductive Issues in Interwar Czechoslovakia. In Hungarian Historical Review 2021, 10(2): 301–327.

Lišková, Kateřina. Sexual Liberation, Socialist Style: Communist Czechoslovakia and the Science of Desire, 1945–1989. Cambridge: Cambridge University Press, 2018.

Olszynsko-Gryn, Jess, Rusterholz, Caroline. Reproductive Politics in Twentieth-Century France and Britain. In Medical History 63 (2019), p. 117–133.

Olšáková, Doubravka, Janáč, Jiří. Kult jednoty: stalinský plán přetvoření přírody v Československu 1948–1964 [The Cult of Unity: Stalin's Plan to Reshape Nature in Czechoslovakia 1948–1964]. Praha: Academia, 2018.

Rákosník, Jakub, Šustrová, Radka, Toward a Population Revolution? The Threat of Extinction and Family Policy in Czechoslovakia 1930s–1950s. In: Journal of Family History 43 (2018), p. 177–193.

Richardson, Angelique, Love and Eugenics in the Late Nineteenth Century: Rational Reproduction and the New Woman. Oxford: Oxford University Press, 2003.

Rusterholz, Caroline, English Women Doctors, Contraception and Family Planning in Transnational Perspective (1930s–70s). In Medical History 63 (2019), p. 153–172.

Shmidt, Victoria, Eugenics and Female Embodiment in Czechoslovak Public Campaigns during the 1960s and 1970s. In Bohemia: A Journal of History and Civilisation in East Central Europe, 58 (2018), p. 109–127.

Szabó, Miloslav. Potraty, dějiny slovenských kultúrnych vojen od Hlinku pod Kuffu [Abortions: A history of Slovak cultural wars from Hlinka to Kuffa]. Bratislava: N Press, 2020.

Usborn, Cornelie. Politics of the Body in Weimar Germany: Women's Reproductive Rights and Duties. Basingstoke: Macmillan Press, 1992.

"Killing of Unborn Children" and "Pornography"
Discourses on Sexuality and Reproductive Rights in Post-war Poland

Michael Zok[1]

Abstract *This article is dedicated to the question of discourses on sexuality in a Catholic, albeit communist, country like Poland. It shows that these discourses could be partially tamed by communist censorship policy, but by the end of communism these discourses broke out completely and continue to dominate discussions until today.*

Introduction

The history of sexuality is often a history of conflicts about its normalization. This also applies to Polish society, where the current conflicts concerning the law on abortion or the one on sex education are obvious. A telltale discourse and notions of "proper sexuality" often shape these clefts. However, the developments in Poland were unique during Communist time and after its downfall: the specifics of the post-war period were that society was recovering from (demographic losses of) German occupation and experienced rapid (forced) industrialization and urbanization. Finally, yet importantly, it was the only Communist country with a strong Catholic Church that challenged the Party's monopoly of power in a "battle over Polish souls". Moreover, during democratic transition in the 1990s, it was one of the few countries worldwide where social liberalisation with regard to sexuality was met with restrictive legal regulations.

Sexuality, also in Polish discourse, has manifold aspects that were at the (multiple) centres of conflicts, thus I will concentrate on just two issues that

---

[1] Deutsches Historisches Institut Warschau (German Historical Institute Warsaw).

were important in the post-war period: the notion of "unborn children" and that of "pornography". In this paper, I will rely on (unpublished) materials produced by different actors, such as the Communist Party, ministries, and Catholic actors and analyse different notions and concepts that rivalled each other. Although especially the legal shifts at the beginning of the 1990s were often described, the change in discourse that took place in the 1980s has been seldomly analysed. I argue that the last decade of Communist rule and particularly the shift of the power balance towards the Catholic Church in this very decade laid ground for those developments that influence Polish discourse until today. To contextualise this, some remarks about the political and social situation have to be given.

"Unborn Children"

Looking back from the current debates about abortion and the existing restrictive law, it is surprising that Poland had liberal laws on abortion for the majority of the 20th century. The reformed penal code of 1932 allowed abortions for medical[2] or criminal[3] reasons. Those regulations were loosened in April 1956 by the parliament, the Sejm.[4] The new law and its preamble[5] put an emphasis on the protection of female reproductive abilities in an effort to prevent illegal abortions in unhygienic surroundings. As in the case of the 1932[6], the new law referred to the foetus as "płód", the Polish translation.

The Communist Party and its MPs saw it necessary to liberalise the law, because of the consequences of the rapid industrialization and urbanization of Poland. Young people were moving to the cities to work in new-built industrial complexes and by migrating, they left social control behind. Press reports about orgies in workers' hotels claimed a growth of sexual contacts resulting in unwanted pregnancies and illegal abortions. Estimates by the Min-

---

2   Meaning in cases of threats to the mother's life or health.
3   Incest, rape, sexual intercourse with a minor.
4   Czajkowska, Dopuszczalności, 2012.
5   https://isap.sejm.gov.pl/isap.nsf/download.xsp/WDU19560120061/O/D19560061.pdf (05.06.2025).
6   https://isap.sejm.gov.pl/isap.nsf/download.xsp/WDU19320600571/O/D19320571.pdf (05.06.2025)

istry of Health suggested that more than 300,000 women were hospitalised after (spontaneous) miscarriages in 1955.[7]

The liberalization led to major conflicts between its supporters and opponents.[8] The main antagonists were the *Polish United Workers' Party* (Polska Zjednoczona Partia Robotnicza, PZPR), the *Women's League* and other state-funded women's organizations as supporters of liberalization. The opponents were led by the Catholic Church and Catholic organizations such as the *Clubs of Catholic Intelligentsia* (KIK). A third party was made up of scientific experts of the state-funded *Society for Conscious Motherhood* (SCM), resp. *Society for Family Planning* (SFP). They were split in their attitude towards the liberal legislation, since on the one hand, they argued that even legal abortions were harmful. However, on the other hand, they saw in legal abortions a "lesser evil" than outlawing these procedures. This would, in their opinion, lead to illegal ones with severe consequences.[9] The question of female sexual self-determination was only of minor priority – even among the supporters.[10]

Opponents of abortions, such as Catholic clerics and conservative politicians, first used the neutral term of "foetus" after the Second World War. In a letter to the Ministry of Labour and Social Welfare in 1946, the author advocated to ban its removals ("spędzanie płodu").[11] This usage of the scientific term lasted until the 1960s, when the then Primate of Poland, Stefan Wyszyński, himself used this term during his meetings with the then Party leader Władysław Gomułka.[12] However, the discourse changed in the course of time and the influence of Catholic thought grew. At its centre was a tendency to humanize the foetus that was called a "conceived child". Opponents of abortion argue(d) that the foetus had "personal rights" that every human being had/has. And, as a follow-up argument, they argued that because it was an autonomous human being with its own rights, a pregnant woman had no right to decide about it.[13]

Those arguments were futile at that time, the liberal law remained, and abortions were a common practice in Peoples' Poland. The official numbers

---

7   Fidelis, Women, 2010, p. 180, 192.
8   Zok, Auseinandersetzungen, 2019.
9   Ignaciuk, Dyskursy, 2014.
10 Zok, Körperpolitik, 2019, p. 144.
11 AAN, URM, 290/0/5/732, f. 1.
12 Raina, Kościół, 1995, vol. 2, p. 10.
13 Zok, Substance, 2021.

gathered by the Ministry of Health and Social Welfare indicated that after an all-time climax in the 1960s, this number declined. At first slowly, then in the second half of the 1980s rapidly.[14] However, there was the problem of undocumented termination of pregnancies in private practices. I discussed this in another article.[15]

From the late 1970s, the PZPR standpoint became more similar to the Church's view. A new party leadership under Edward Gierek had come into power at the beginning of the decade and ended open confrontation with the Church. Instead, a rapprochement of both sides took place. This led to an upvaluation of family[16] and therefore to a redefinition of women as "mothers [and] workers" (in this order of priority)[17] in official propaganda. Another factor for a conservative turn was the perception of demographic "problems": from the 1960s, the average number of children per family was decreasing and especially the Church's hierarchy saw potential "threats to the biological substance of the nation". Therefore, it addressed these concerns to the government and Party leadership twice in the 1970s.[18] While the party leadership under Władysław Gomułka did not react to the concerns, the Gierek administration at least partly saw the same problems. For example, the PZPR Administrative Department called the dominating "tendency of families with only one or two children […] alarming".[19]

Economic problems were a characteristic part of Communist society. However, due to investments planned by the Gierek administration that were financed by Western banks and institutions indebtedness rose. Economic imbalances and crises erupted at the beginning of the 1980s. Women were affected most prominently, since they were either the ones who had to stand in line for goods, which led to the emergence of a "society of queues".[20] Alternatively, the regime tried to "rescue socialist economy" by removing women from the labour market.[21]

Another factor was the growing dominance of Catholic Social Thought in public discourse. The reason for this was the Church's role as mediator between

---

14 AAN, MZiOS, 1939/20/27, f. 1.
15 Zok, Substance, 2021, p. 357–359.
16 Stegmann, Aufwertung, 2005.
17 Stańczak-Wiślicz et. al., Kobiety, 2020, p. 139.
18 Zok, Auseinandersetzungen, 2019.
19 AAN, KC PZPR, 1354/XI-970, f. 122.
20 Mazurek, Społeczeństwo kolejki, 2010.
21 Zok, Policies, 2022.

"state/Party" and *"Solidarność*/society" during the crisis on the one hand. On the other, the election of Karol Wojtyła as Pope John Paul II in 1978 and his journeys to Poland, which strengthened the Church's position. Thus, Catholic notions, such as the "unborn child" began to dominate public discourse. Its "protection" from the "moment of conception" onwards was the common goal of Church, laypersons, KIK, and, after 1989, the newly established right-wing parties.[22]

The first draft bill was introduced in the Sejm in February 1989 (before the semi-free elections of June that ended the PZPR's political monopoly) by a group of (mostly male) MPs, some of them from PZPR. The Party members were surprised by the speed it developed, since the first reading was unexpectedly carried out before the June elections.[23] After the elections, *Solidarność* dominated the re-established Upper House, the Senate, and could initiate own projects. One of the first initiatives was a very restrictive draft bill based upon a concept by a Church commission. It would have not only outlawed abortions, but also contraceptives. It was stopped by the Sejm.[24]

However, right-wing and centrist parties dominated the Sejm after the first free elections of 1991 and passed a new restrictive law on January 7, 1993, often called a 'compromise': although it did not outlaw abortions in general, it denied women the right to have one on request. Instead, the regulations of the 1932 law plus embryo-pathologic reasons returned. Only in its last section the term "foetus" (płód) can be found. Although entitled "Law on Family Planning, the Protection of the Foetus, and the Circumstances of the Permissibility of Termination of Pregnancies", in all other than the above-mentioned section, it refers to the foetus as "conceived child",[25] thus using a term taken from the Catholic discourse.

The draft bill had been heavily criticised, because of the turbulences of economic transformation and mass unemployment among women and surveys showed that a majority of interviewees did not want the law tightened. One MP even warned of damages to the young Polish democracy, if the parliament would vote for a law that did not have the approval of society. A proposed solu-

---

22 Zok, Kompromiss, 2021.
23 AAN, KC PZPR, 1354/XXV-28, n.p.
24 Kulczycki, Policy, 1995, p. 483–484.
25 https://isap.sejm.gov.pl/isap.nsf/download.xsp/WDU19930170078/O/D19930078.pdf (05.06.2025).

tion through a referendum about the penalisation of abortions was rejected by the right-wing majority in parliament.[26]

However, the 'compromise' did not end conflicts. Attempts to either loosen or tighten the law have been made since then,[27] but there have been no significant changes. Because of the Constitutional Tribunal's verdict deeming the embryo-pathological indication as unconstitutional in October 2020, the issue entered the public once again.

"Pornography"

Abortion was one issue in which divergences between Church teachings and (restrictive) legal regulations on the one side and liberalising trends and actual behaviour on the other were visible. Another issue was nudity in media, often either regarded as "pornography" or "ars erotica". Almost at the same time as abortion, nudity became a bone of contention between liberals and traditionalists. Besides the abovementioned liberalization of reproductive rights, the "Political Thaw" of 1956 also led to a liberalization in the media. However, these developments led to mixed reactions.

At first, nudity in media was met by critique. One female Party member in Katowice criticized the newspaper *Dziennik Zachodni* not only because it published pictures of naked women next to news about politicians from PZPR. Moreover, it had also initiated a contest for women to get naked in public, but the (male) journalists could not find enough women who were ready to take part. The female functionary was furious at the journalists for the idea of this contest and condemned this "sztryktis" [sic!] as *"the* disgrace to the 20th century".[28]

However, the 1970s showed liberal attitudes towards nudity and sexuality, since the new Gierek administration tried to open Poland for Western investments. New forms of discourse and behaviour appeared, e.g. the emergence of well-situated female sex workers concentrating on foreign businessmen which changed the discourse on prostitution.[29]

---

26 AAN, ZChN, 2410/6, n.p.
27 Ignaciuk, Abortion, 2007.
28 AAN, KC PZPR, 237/VII-2946, f. 272–273.
29 Dobrowolska, Prostytucja, 2020.

These liberalising trends were most openly visible in mass-media, especially television. Some viewers were outraged at "obscene" scenes and accused television of "demoralizing the youth". To prove this accusation, a female viewer from Poznań claimed that

> these films encourage [young people] to have sexual intercourse leading to venereal diseases. These directors and actors do not know any shame. Almost every film has disgusting love scenes and children and young people watch them. The reviews [in newspapers] describe the movies as psychodramas and tragedies, but later, one sees that these films consist only of naked people and love scenes.[30]

Nonetheless, some viewers were in favour of nudity in television. As one wrote:

> We want to see more sex. [...] Does the leadership of television worry about our morals? Do we live in times of petty bourgeois dishonest morals, or do we want to return to them? [I am convinced that] open, frivolous scenes that show sex will not harm socialism [...].[31]

Another one argued that "[a]s if nudity and sex would endanger socialist education and conscience! Or [is the leadership of television] afraid of old grannies and sullen moralists?"[32]

However, these "sullen moralists" were also among members of PZPR. The authors of a letter to the leader of the Department of Propaganda, Press and Publications described themselves as devoted communists and saw in the showing of naked women – they referred to a figure photography of a young woman from different angles – a "fascist capitalist diversion" that was dominating People's Poland by means of "fashion [and] pornography". The authors advocated a ban on such pictures and punishment for those who had allowed the circulation of the pictures.[33]

Critique about publications that were seen as potentially "demoralizing" the youth and family life also came from Catholic clerics and laypersons. In one of the aforementioned aide-memoirs from the 1970s, the Bishops' Conference demanded the introduction of several bans including "pornography" and

---

30   Zok, Darstellung, 2015, p. 144–145.
31   Zok, Darstellung, 2015, p. 145.
32   Zok, Darstellung, 2015, p. 146.
33   AAN, KC PZPR, 1354/XXXII-55.

all publications that propagated "sexual freedom" as well as advertisement for contraception[34] – which was legal in Poland, but was in general of low quality.[35] The Bishops' Conference saw a problem for "a proper education" of children because of the perceived omnipresence of "pornography".[36]

However, the ambivalence about the notion of what was "appropriate" and what was not remained – as well as the lack of a clear definition and differentiation between "pornography" and "ars erotica". While for conservatives, nudity in general was associated with sexuality (and sexual desire), liberals underlined the aesthetics and beauty of the human (mostly female) nude body. This irreconcilable view on nudity could not been overcome until the end of Communist reign, thus a law banning "pornography" was never established, since there was no common definition.

Despite that, the Censorship Office was highly involved in the question of what "pornography" was. It saw its tasks in "protecting good mores" and hindering a "wave of pornography". Especially publications from abroad that showed or described "sexual relationships in a drastic manner" were excluded from distribution. This included different media, such as *Easy Rider*, gazettes with salacious content like *Playboy*, *St. Pauli Nachrichten* as well as some numbers of the German youth magazine *Bravo*. Furthermore, the censors hindered the publication of works by Crumbe and de Sade, and the fictional *Fanny Hill*. In their opinion, the connection of sexuality, violence, and drugs was one reason to confiscate these publications.[37]

However, a detailed instruction on how to recognize and categorize some publications as "pornographic" was lacking.[38] In 1977, one of the first studies about this problem was published and the author referred also to the problem of a lacking definition. Summing up his research, he even discussed a legalization of "pornography", because of its ambivalent character and the problems the jurisdiction had to come to terms with it.[39] A secret report from the 1980s stated that the majority of society accepted "erotic content", but only if

---

34 AAN, UdsW, 1587/125/120, f. 50.
35 Ignaciuk, Dyskursy, 2014.
36 AAN, UdsW, 1587/136/163, f. 91.
37 AAN, GUKPPiW, 1102/3742, f. 112–113.
38 AAN, GUKPPiW, 1102/3715, f. 9–10.
39 Filar, Pornografia, 1977, p. 5, 162–164.

the "confrontation with nudity was voluntarily". Thus, the report summed up, nudity was not the criterion to ban a publication for being "pornographic".[40]

Beata Łaciak argues that if there was a "sexual revolution" in Poland, it was during the last decades of Communism,[41] when (liberal) sexologists emphasized the importance of "(female) lust" for a relationship and a happy sex life.[42] Surveys from the 1980s supported this and showed that society was (and probably still is) more liberal than politicians. Its results indicated that almost two-thirds of interviewees were against a ban on pornography and in favour of an unhindered distribution of magazines and screening of films in selected theatres. The authors of the report summed up that this showed the youth's interest in sexuality.[43]

The conflict about nudity in mass media intensified at the beginning of the transformation period. The introduction of a market economy after 1989 led to a rise of magazines that showed naked women on covers and could be bought at every kiosk. A "crusade against pornography" was proclaimed in which KIK took part. E.g., its Gdansk branch criticized the decision to screen "Moulin Rouge" on New Year's Eve 1989/90, and it attacked Polish Radio for broadcasting a feature about the novel "Emmanuelle". The critics argued that its "pornographic character is widely known". Members of KIK were also dissatisfied with the changes during transition and angrily summed up that morals had not changed for the better. Instead, they perceived an ongoing "assault on our traditions" and stated that such "smut" should not be protected by freedom of speech or art.[44]

However, some of the film reviewers did not agree, or at least didn't share the view of it being "pornographic". In his review of "Emmanuelle", Marek Antczak stated in *Głos Pasłęka*, in November 1992:

> Old trash. Its broadcast led to some agitation. Especially within circles of people who, at least with regard to their age, should have seen a naked woman and experienced "such things". One MP said he did not even watch the movie, but he had seen pictures, [however] he was convinced that it was pornography, [although instead of watching] he went to religious course.[45]

---

40   AAN, KC PZPR, 1354/XXXVII-39, f. 21–22.
41   Laciak, Gender, 1996, p. 38.
42   Kościańska, Sex, 2016.
43   AAN, KC PZPR, 1354/LII-211, n.p.
44   AAN, KIK, 2212/11, n.p.
45   AAN, UD, 2956/9, n.p.

The Gdansk KIK branch was convinced that "pornographic" material could lead to a change in human sexuality and behaviour. Its members argued (counter-factually) that the increase of these magazines and the "erotization of social life" had led to more teenager pregnancies and, in general, to more abortions – although the numbers of registered abortions were falling since the late 1980s. The government's silence on this problem was interpreted as acceptance. The Club members stated that "although we respect the freedom of speech, we expect that the rights of persons, families and of the society to protect its basic moral norms" should be enforced and condemned all sorts of "hedonism".[46] This examples shows the new factors that influenced the discourse on "pornography" at the beginning of the 1990s: on the one hand, economic transformation and introduction of free entrepreneurship leading to an emerging 'sex industry', on the other hand, discussions about the (wished) "traditionalist" and "Catholic" character of state and nation.

Additionally, the end of (Communist) censorship showed an increased importance of interplay with the public. Not only the KIK wrote "open" letters complaining about morals, but also clerics, such as Reverend Stanisław Wlazło. Referring to unknown "experts", he claimed that the bad moral state of youth was "even more dangerous than the economic situation". And, he was angry because the takeover of power by *Solidarność* had led to more, instead of less, "pornography" on television. In his eyes, especially young people would lose their "natural" prudence and the youngest would become "victims" to "sexualism". Finally, he was convinced that "pornography will not rescue the fatherland, [instead] it is the enemy of all ideals."[47]

This clash was also observable in the political discourse. The *Polish Forum of Christian-Democrats* stated in February 1993, that "democracy cannot apply to all levels" of society and demoralization by "pornography" had to be prevented by outlawing it.[48] This notion was rejected by centrist and left-wing parties, such as the *Social Democracy of the Republic Poland*, the successor of PZPR. One of its MPs reported to his colleagues the discussions at a meeting of the Sejm's Commission on Family. Latter had debated about banning "pornography", but also all sorts of "erotica" in an effort to prevent sexual harassment of children. Disagreeing with this, the leftist MP argued that forbidding all sorts of nudity

---

46   AAN, KIK, 2212/11, n.p.
47   AAN, KIK, 2212/11, n.p.
48   AAN, PFChD, 2093/8, f. 113.

was "nonsensical", because it would "prohibit adults something that existed in normal life".[49]

Conclusion

This paper has exemplified conflicts on sexuality and related issues with the examples, abortions and pornography. It has shown that liberalising trends in society were often not acknowledged by political, and even more by religious elites. However, their efforts to control and influence sexual behaviour of individuals as well as the collective were unfruitful. Neither did the outlawing of abortion in 1993 make it disappear, nor did the discussions about "pornography" raise "sexual knowledge" and "protected minors". Instead, both examples show how conservatives tried to implement their agenda in legal form, since individual behaviour remained uninfluenced on a large scale, despite all efforts to propagate Church teachings. Abortions remained a common experience of women during Communism and nudity was (and still is) pictured in mass media.

The papers shows also one important observation: the instrumentalisation of sex-related issues in a discourse about nation as an imagined "ideal Catholic community". Irreconcilable concepts like the humanisation of the foetus vs. (female) reproductive rights and nudity as "pornography" vs. "ars erotica" dominate(d) Polish discourse. Moreover, surveys from late Communism and early transformation indicated that a majority of Polish citizens wanted nudity in mass media, a liberal law on abortion, and in general a liberal stance on sexuality. However, the decisions by the new right-wing political elite aimed at a restrictive attitude towards sexuality and the then climaxing conflicts are still a part of the Polish discourse, since their underlying fundamental problems – a "low sex culture", unwanted pregnancies and the instrumentalisation of those issues in the political discourse – have not been solved until today. Further studies about the actors, their motives and strategies as well as about continuities and fractions will be needed.

---

49   AAN, SLD RK, 2590/19/6, n.p.

Bibliography

Archival Sources

Archiwum Akt Nowych w Warszawie (Archive of Modern Records Warsaw)
Główny Urząd Kontroli Prasy, Publikacji i Widowisk
Komitet Centralny Polskiej Zjednoczonej Partii Robotniczej
Klub Inteligencji Katolickiej
Ministerstwo Zdrowia i Opieki Społecznej
Porozumienie Centrum
Polskie Forum Chrześcijańsko-Demokratyczne
Sojusz Lewicy Demokratycznej. Rada Krajowa
Unia Demokratyczna
Urząd do spraw Wyznań
Urząd Rady Ministrów

Literature

CZAJKOWSKA, Aleksandra, O dopuszczalności przerywania ciąży. Ustawa z dnia 27 kwietnia 1956 r. i towarzyszące jej dyskusje, in: Zawodowe dziewczyny. prostytucja i praca seksualna w PRL, Warszawa, Wydawnictwo Krytyki Politycznej, Seria Historyczna, 2020, p. 99–186.

DOBROWOLSKA, Anna, Zawodowe dziewczyny. prostytucja i praca seksualna w PRL, Warszawa, Wydawnictwo Krytyki Politycznej, Seria Historyczna, 2020.

FIDELIS, Małgorzata, Women, communism, and industrialization in postwar Poland, Cambridge [etc.], Cambridge University Press, 2010.

FILAR, Marian, Pornografia. Studium z dziedziny polityki kryminalnej, Toruń, Wydaw. UMK, Rozprawy, Uniwersytet Mikołaja Kopernika, 1977.

IGNACIUK, Agata, Abortion Debate in Poland and its Representation in Press, Łódź 2007.

IGNACIUK, Agata, Ten szkodliwy zabieg. Dyskursy na temat aborcji w publikacjach Towarzystwa Świadomego Macierzyństwa, Towarzystwa Planowania Rodziny (1956–1980) In: Zeszyty Etnologii Wrocławskiej 20.1 (2014), p. 75–97.

KOŚCIAŃSKA, Agnieszka, Sex on equal terms? Polish sexology on women's emancipation and 'good sex' from the 1970s to the present, in: Sexualities, 19 (2016), p. 236–256.

Kulczycki, Andrzej, Abortion Policy in Postcommunist Europe. The Conflict in Poland, (1995), p. 471–505.

Laciak, B., Sex, gender and body in Polish democracy in the making, 10 (1996), p. 37–51.

Mazurek, Małgorzata 1979-, Społeczeństwo kolejki, o doświadczeniach niedoboru 1945–1989, Warszawa, Wydawn. Trio [u.a.], 2010.

Raina, Peter, Kościół w PRL, vol. 2, Lata 1960 – 74, Poznań, 1995.

Stańczak-Wiślicz, Katarzyna, Perkowski, Piotr, Fidelis, Małgorzata & Klich-Kluczewska, Barbara, Kobiety w Polsce 1945–1989. Nowoczesność, równouprawnienie, komunizm, Kraków, Towarzystwo Autorów i Wydawców Prac Naukowych Universitas, 2020.

Stegmann, Natali, Die Aufwertung der Familie in der Volksrepublik Polen der siebziger Jahre, Jahrbücher für Geschichte Osteuropas (2005), p. 526–544.

Zok Michael, "To Maintain the Biological Substance of the Polish Nation". Reproductive Rights as an Area of Conflict in Poland, in: Hungarian Historical Review, 10 (2021), p. 357–381.

Zok, Michael, (K)Ein Kompromiss? Der Konflikt um die Neuregulierung des Schwangerschaftsabbruchs in Polen in den 1980er/1990er Jahren, in: Ariadne. Forum für Frauen- und Geschlechtergeschichte, 77 (2021), p. 164–181.

Zok, Michael, Wider die "angeborene und nationale Mission der Frau"? Gesellschaftliche Auseinandersetzungen um Abtreibungen in Polen seit der Entstalinisierung, in: Zeitschrift für Ostmitteleuropa-Forschung, 68 (2019), p. 249–278.

Zok, Michael, Körperpolitik, (staatstragender) Katholizismus und (De-)Säkularisierung im 20. Jahrhundert. Auseinandersetzungen um Reproduktionsrechte in Irland und Polen, in: Body Politics, 7 (2019), p. 123–158.

Zok, Michael, Die Darstellung der Judenvernichtung in Film, Fernsehen und politischer Publizistik der Volksrepublik Polen 1968–1989, Marburg, Verlag Herder-Institut, Studien zur Ostmitteleuropaforschung, 2015.

Authors

Aleksandra Jakóbczyk-Gola (Post-doctoral degree) works as a curator of historical exhibitions in Polish History Museum and as a researcher at Artes Liberales Faculty at University of Warsaw.

Hadrian Ciechanowski (dr hab.) is an assistant professor at Nicolaus Copernicus University in Toruń (Poland). He teaches digital humanities in history studies. His second interests are the history of bureaucracy, public registers, and genealogy.

Heidi Hein-Kircher (Prof. Dr.) is director of Martin Opitz Library and Professor for German Culture and History in Eastern Europe at Ruhr University Bochum. Her interests lay among others in the history of family planning and anti-feminists movements in East Central Europe.

Elisa-Maria Hiemer (PhD) has been working a postdoctoral researcher in the ERC-Project "Democratising the Family?" at FU Berlin since December 2024. She authored the monograph "Maternity in Times of Crisis? Voices From Interwar Poland on Sexuality and Abortion" (CEU Press, forthcoming).

Stefan Jehne (PhD-candidate), works as a research assistant on the project to redesign the permanent exhibition at the Hadamar Memorial.

Veronika Lacinová Najmanová (PhD) is teaching gender history at University of Pardubice. She researches and writes about birth control in the 20th century.

Wiebke Lisner (PD Dr.) is teaching history of medicine at Hannover Medical School. She researches and writes about midwifery in the 20th century and also about pandemic planning since the 1990s.

Katarzyna Pekacka-Falkowska (PhD) teaches history of medicine and allied sciences at Poznan University of Medical Sciences.

Małgorzata Radkiewicz (PhD) is an Assistant Professor at the Institute of Audio Visual Arts at the University of Krakow. Her research interests and publications focus on gender representation in film and media as well as on much wider category of cultural identity.

Dr. **Tim Rütten** is a research assistant for the history of the early modern period at the Institute of History.

Izabela Spielvogel (PhD) represents the fields of medical and health sciences, history of medicine and ethnomedicine. She obtained her habilitation in medical and health services in 2025.

Marcin Wilk (MA Polish Studies, Jagiellonian University), is an PhD candidate at Institute of History, Polish Academy of Sciences.

Michael Zok is a research assistant at German Historical Institute Warsaw.